PHYLLOTACTIC PATTERNS
A MULTIDISCIPLINARY APPROACH

PHYLLOTACTIC PATTERNS
A MULTIDISCIPLINARY APPROACH

DENIS BARABÉ
Université de Montréal, Canada

CHRISTIAN LACROIX
University of Prince Edward Island, Canada

World Scientific

NEW JERSEY · LONDON · SINGAPORE · BEIJING · SHANGHAI · HONG KONG · TAIPEI · CHENNAI · TOKYO

Published by

World Scientific Publishing Co. Pte. Ltd.

5 Toh Tuck Link, Singapore 596224

USA office: 27 Warren Street, Suite 401-402, Hackensack, NJ 07601

UK office: 57 Shelton Street, Covent Garden, London WC2H 9HE

Library of Congress Cataloging-in-Publication Data
Names: Barabé, Denis, 1951–   author. | Lacroix, Christian (Biology Professor), author.
Title: Phyllotactic patterns : a multidisciplinary approach / Denis Barabé, Christian Lacroix.
Description: Hackensack, NJ : World Scientific, [2020] |
    Includes bibliographical references and index.
Identifiers: LCCN 2020001688 | ISBN 9789811211003 (hardcover) |
    ISBN 9789811211010 (ebook)
Subjects: LCSH: Plant morphology. | Biomathematics.
Classification: LCC QK641 .B36 2020 | DDC 571.3/2--dc23
LC record available at https://lccn.loc.gov/2020001688

British Library Cataloguing-in-Publication Data
A catalogue record for this book is available from the British Library.

For any available supplementary material, please visit
https://www.worldscientific.com/worldscibooks/10.1142/11571#t=suppl

Contents

Acknowledgements

We thank our colleagues and friends who have provided feedback on different portions of this book: Yann Guédon, Xiaofeng Yin, Bernard Jeune, and Patrick Shipman. We also thank the anonymous reviewers for their helpful and insightful comments. Their suggestions greatly improved this manuscript. We accept full responsibility for any errors or omissions that may appear in the final version of this book.

A special thank you to our graphic artist Dave Goulet who reproduced many of the excellent figures we selected from the phyllotaxis literature. We also appreciate the tremendous effort of an anonymous editor at Editage who reviewed the manuscript for style and grammar. Finally, we thank World Scientific Publishing for supporting our idea and making this book project a reality.

We are grateful to the following organisations and institutions for their in-kind or financial support: The University of Prince Edward Island, Institut de recherche en biologie végétale (Jardin botanique de Montréal, Université de Montréal), Natural Sciences and Engineering Research Council of Canada (NSERC), and the Canadian Botanical Association/L'Association botanique du Canada.

The inspiration for this book comes from our interest in patterns in plant and our respective research careers that have touched on different aspects of phyllotaxis. We credit our mentors Roger V. Jean, Rolf Sattler, Joachim Vieth, and Usher Posluszny for encouraging us to pursue our research interests in plant development. Finally, a very special thanks to our families for their unwavering support.

Introduction

The manifestation and analysis of the regular patterns of leaf arrangement on plants have been documented since the time of Theophrastus (c. 371–287 BC). The term phyllotaxis, coined by the German botanist Karl Fr. Schimper in the 19th century, originates from the Greek terms φύλλον (phullon, meaning leaf) and τάξισ (taxis, meaning arrangement) and refers to the disposition of leaves on the stem (Loiseau, 1969). Barlow (1994) distinguished the terms phyllotaxy and phyllotaxis. According to this author, phyllotaxy refers to the 'state' of the arrangement of leaves on a shoot system and phyllotaxis is the process that leads to the positioning of these structures on the shoot apical meristem (SAM). Phyllotaxis was defined more comprehensively by Jean (1994) as the description, characterisation, and generation of the patterns made by similar elements (e.g. leaves, floral parts), and the study of the growth processes leading to their formation. Initially, phyllotactic analyses were conducted mainly on mature plants but gradually expanded to include the earliest stages of the organs that develop at the periphery of the SAM (i.e. leaves and flowers), where specific phyllotactic patterns are established. Throughout history, several scientists have examined this phenomenon from a variety of perspectives ranging from morphological, mathematical, and experimental studies to the development of models that take into consideration mechanical, biochemical, and molecular aspects of development (Table I.1).

Table I.1.   Brief history of phyllotaxis and pioneering contributions to the discipline

Date	Authors	Contribution
370–285 BC	Theophrastus	*Enquiry into Plants* (1916 edition); mentions leaves in regular series
23–79 AD	Pliny the Elder	*Natural History* (1856 edition); described the position of leaves around the stem
1202	Leonardo Fibonacci	*Liber Abaci* (Book of Calculation); used the mathematical series $\langle 1, 1, 2, 3, 5, \ldots, F(k), F(k+1)\rangle$ to describe population growth of rabbits
1452–1519	Leonardo Da Vinci	In his notebook, anticipated Bonnet's observations regarding the spiral arrangement of leaves (Bonnet's *quincunx*); noted in Adler (1974)
1611	Johannes Kepler	*De nive sexangula* (On the Six-Cornered Snowflake); noted the importance and prevalence of Fibonacci numbers in plants
1754	Charles Bonnet	*Recherches sur l'usage des feuilles dans les plantes*; initiated observational phyllotaxis by recognising different phyllotactic patterns
1830	Karl Friedrich Schimper	Defined genetic spiral, divergence angle, and parastichies during early days of quantitative phyllotaxy
1831, 1835	Alexander Braun	Applied Schimper's concepts in the study of parastichies and their relation to Fibonacci numbers
1837, 1839	Auguste and Louis Bravais	Represented leaf arrangement as points on a cylindrical lattice and derived a formula for approximating the divergence angle based on the number of parastichies
1848	Gaspard Thémistocle Lestiboudois	Studied anatomy and linked phyllotactic patterns to the phenomena of ramification and branching

Table I.1.   (*Continued*)

Date	Authors	Contribution
1868	Wilhelm Hofmeister	Investigated phyllotaxis at the SAM level; formulated Hofmeister's rule: primordia originate the furthest away from the base of existing leaves
1872	Peter Guthrie Tait	Like Bravais and Bravais, developed a method to calculate the divergence angle based on the number of parastichies
1873	Hubert Airy	Proposed a functional approach to phyllotaxis based on leaf packing and economy of space in buds
1875	Julius Ritter von Wiesner	Provided a functional explanation of phyllotaxis in relation to ecology and maximisation of leaf exposure to light
1878, 1883	Simon Schwendener	Provided mechanical and causal explanations of phyllotaxy; developed a theory of contact pressure between primordia to explain leaf arrangements; established a mathematical relation between divergence angle and radius of contiguous units (i.e. lateral organs) on an open cylinder
1904	Arthur Harry Church	Used planar representations of phyllotactic patterns where leaves are represented as points inside a disc; also developed the idea of periodic pulses or rhythm to explain phyllotaxis
1907	Gerrit van Iterson	Modelled the packing of different-sized lateral organs along the SAM; showed how phyllotactic patterns change by varying the ratio of the primordium size in relation to the size of the SAM

(*Continued*)

Table I.1.   *(Continued)*

Date	Authors	Contribution
1913	J. C. Schoute	Attributed phyllotaxis to a chemical inhibitor: an inhibitor is produced by a leaf primordium, thereby preventing the formation of another primordium in its vicinity
1931	Mary Snow and Robert Snow	Proposed the theory of first available space: a primordium forms in a specific area of the apex where there is adequate space for its initiation
1948, 1951	F. J. Richards	Developed the notion of plastochrone ratio and described the relationship between divergence angle, plastochrone ratio, and form of the apex
1948	Lucien Plantefol	Proposed the theory of foliar helices: this idea is linked to the concept of generative centres operating at the leading end of the helices at the SAM level
c. 1952	Alan Turing	Used a reaction–diffusion system to explain phyllotaxis (in Swinton, 2005)[a]
1973	Ralph O. Erickson	Applied van Iterson's principles to different biological structures
1974	Irvin Adler	Developed a model of contact pressure in phyllotaxis based on the maximisation of the minimum distance between primordia; first to formulate the Fundamental Theorem of Phyllotaxis

[a]Turing's seminal work on morphogenesis (1952) does not include phyllotaxis. His work on phyllotaxis was included many years after his death in *Collected Works of A.M. Turing* (Turing, 1992). In this paper, he applies a reaction–diffusion equation to the problem of phyllotaxis. Turing's work on the chemical theory of phyllotaxis was analysed in detail by Swinton (2005) and Rueda-Contreras and Aragón (2014).

Table I.1. *(Continued)*

Date	Authors	Contribution
1988	Roger Jean	Presented a general form of the Fundamental Theorem of Phyllotaxis
1991	Leonid Levitov	Proposed an energetic approach to phyllotaxis; new primordia form where the repulsive energy from existing primordia is lowest
1992	Stéphane Douady and Yves Couder	Created a paradigm for elaborating a dynamical system of phyllotaxis based on repulsive energy; experimentally showed that purely physical mechanisms play roles in the elaboration of phyllotactic patterns
1992	Paul Green	Proposed a biophysical principle to explain phyllotaxis by making an analogy to bump formation on a sheet experiencing bucking pressure
2000, 2003	Didier Reinhardt *et al.*	Experimentally showed the central role of auxin in morphogenesis and phyllotaxis
2005	Patrick Shipman and Alan Newell	Elaborated a biophysical energy-minimising buckling model to generate phyllotactic patterns and planforms
2006	Richard Smith *et al.*	Performed a computer simulation of phyllotactic patterns based on active auxin transport at the cellular level
2006	Henrik Jönsson *et al.*	Developed simulations of phyllotaxis with a focus on PIN1 dynamics and auxin transport and introduced cell and wall compartments in their model
2008	Alan Newell *et al.*	Provided the first iteration of a model including mechanical buckling of the plant tunica and biochemical processes involving auxin flux

(Continued)

Table I.1. *(Continued)*

Date	Authors	Contribution
2012	Vincent Mirabet *et al.*	Incorporated stochastic elements in a dynamical model based on inhibitory fields
2016	Yann Refahi *et al.*	Developed a stochastic model of phyllotaxis based on biological and physical parameters

Note: For a more comprehensive description of contributions to the development of phyllotaxis, see Loiseau (1969), Adler (1974), Rutishauser (1981), Jean (1994), Adler *et al.* (1997), and Pennybacker *et al.* (2015).

Three main types of regular phyllotactic patterns can be distinguished in plants based on morphological characteristics (Fig. I. 1): spiral (leaves inserted in a spiral around the stem), whorled (two or more leaves inserted at the same node), and distichous (leaves inserted alternately in two rows on opposite sides of the stem). Atypical patterns, in which leaves are distributed more or less randomly on the stem, also exist. Although such patterns are not frequently observed in nature, they can appear in artificially produced plants, as is the case with *Arabidopsis thaliana* (Mähönen *et al.*, 2006) and *Zea* spp. (Itoh *et al.*, 2000) mutants. These patterns add new perspectives and hypotheses, but also lead to problems in the development of theoretical phyllotactic models. Other (less common) types of phyllotactic patterns are variations of the basic morphological types listed above. For example, the spirodistichous pattern is essentially a distichous arrangement where the two rows of leaves on either side of the stem twist to form an arrangement that looks like a spiral staircase.

Throughout the history of phyllotaxis, several parameters have emerged to characterise and quantify the displayed patterns, thereby establishing a common basis for their study. A detailed

Fig. I. 1. Examples of phyllotactic patterns based on morphology. **(A)** Spiral pattern of leaves in *Vriesea gigantea* (Bromeliaceae). **(B)** Decussate pattern in *Coleus* sp. (a cultivar of *Plectranthus cutellarioides*). Sequential numbers indicate younger to older leaves. **(C)** Distichous leaf pattern in *Begonia* 'Honeysuckle' (cane-like group). **(D)** Spiral pattern in *Helianthus* sp. capitulum in which contiguous florets are arranged in counterclockwise (34) and clockwise (51) spirals; representative spirals are highlighted by white lines.

survey of these parameters and their application in phyllotaxis can be found in the chapter "Phyllotactic Parameters."

The generation of symmetrically consistent patterns displayed by similar elements is a general and well-documented phenomenon observed in diverse biological or physical structures (Jean, 1994; Jean and Barabé, 1998a, 1998b; Lieber, 1998; Boeyens, 2003). For example, phyllotactic-like patterns in terms of element layout have been described in the energy spectra of quasiperiodic systems (Macia *et al.*, 1994), carbon nanotubes (Wildoer *et al.*, 1998), viruses (Marzec, 1998, 1999a, 1999b), proteins (Jean, 1994, p. 219), *Escherichia coli* bacterial cultures (Budrene and Berg, 1991), a mechanically annealed experimental ⟨⟨magnetic cactus⟩⟩ (Nisoli *et al.*, 2010), and in the bubbles emerging periodically at the surface of a viscous liquid (Yoshikawa *et al.*, 2010). The mathematical regularity of these patterns has prompted many studies in nearly all scientific disciplines. However, from the botanical perspective, phyllotactic patterns remain unresolved at the fundamental level, although they have been studied and explained in detail from several other perspectives.

Given the diversity of phyllotactic patterns in plants, it is not surprising that botanists, mathematicians, and physicists have tried to uncover the rules governing their appearance over the past 150 years (Jean, 1994; Adler *et al.*, 1997). The relationship between botany and mathematics was established a long time ago and was aptly commented on by the anatomist Nehemiah Grew, who stated that "from the contemplation of plants, Men might first be invited to mathematical inquiries" in 1682. It is the precise organisation of appendicular elements (flowers or leaves) presenting rotational symmetry on a cylindrical or discoid structure, such as the well-known sunflower capitulum, which highlighted the idea of natural forms governed by mathematical rules. Hervé Le Guyader (1988) poetically described this fascination for numbers and regularities in phyllotaxis as "la phyllotaxie ou le rêve du cristal vivant" (phyllotaxis or the dream of the living crystal).

A truly synergetic collaboration between mathematicians and botanists began with brothers Auguste Bravais (crystallographer) and Louis Bravais (physicist–botanist), who made a remarkable contribution to the discipline at the beginning of the 19th century by providing the fundamentals for a descriptive mathematical approach to phyllotaxis. This mathematical approach was followed by botanists Schwendener (1878, 1883) and van Iterson (1907), who proposed a causal mechanistic explanation for phyllotaxis. As noted by Okabe (2016), in the 1930s, several German botanists analysed the frequent occurrence of Fibonacci numbers and the variation of divergence angles in different groups of plants from an empirical point of view. The pioneering experimental work of Snow and Snow (1931) established the basis for explaining the influence of pre-existing primordia on the location of new incipient primordia at the level of the SAM, thereby providing a new perspective and a deeper understanding of the origin of phyllotactic patterns. This paved the way to a variety of physiological and developmental experiments to study phyllotaxis.

This complementarity between developmental plant morphology and mathematics has been extensively examined with relatively successful outcomes. Whereas some studies focussed on the development of models to explain the regularity and recurrence of phyllotactic patterns, others attempted to unravel the biochemical and biophysical processes involved in the generation and display of phyllotactic patterns. It is therefore not surprising that experimentalists and theorists have followed one another in proposing different explanations for phyllotactic patterns (Adler *et al.*, 1997), progressing from almost entirely descriptive to more explanatory models.

Phyllotaxis is one of the most exciting biological phenomena because it is at the interface of many disciplines such as molecular biology, mathematics, physics, biochemistry, and botany. Given its importance in botany and theoretical biology, research in phyllotaxis has always alternated between experimental and theoretical studies, the two complementary approaches used for examining phyllotactic patterns. The studies aiming to understand phyllotactic patterns comprise two main questions: What is the mathematical basis of phyllotactic patterns? How are phyllotactic patterns generated? The first question refers to the geometric description of phyllotactic patterns by using well-defined quantitative parameters. The second deals with the biological, chemical, and physical explanations of phyllotactic patterns, their emergence, initiation, and maintenance.

The SAM, which is the organising centre where phyllotactic patterns are initiated and perpetuated, has different forms, sizes, and complexity. Are there 'universal' principles governing the mode of initiation of plant elements that can be described mathematically or genetically? More specifically, can the initiation of new primordia in specific, regular, and repeatable patterns be described and reproduced using mathematical equations or models? Advances in molecular biology and computer science have provided opportunities to develop more realistic and integrated models to explain the dynamics of organ initiation and the establishment of phyllotactic patterns at the level of the SAM (Prusinkiewicz and Runions, 2012).

Recent molecular biology studies have provided new insights into the genetic and biochemical processes involved in the structural organisation of the SAM and generation of phyllotactic patterns. More specifically, the role of genes linked with the production of auxin was elucidated and interpreted in the context of the overall organisation of the SAM. Based on these results, plausible models of phyllotaxis involving the diffusion of auxin and interaction between genes responsible for the regulation of this hormone have been proposed (e.g. Smith *et al.*, 2006b). This ground-breaking development has provided new information on the mechanisms responsible for maintaining the highly precise patterns of organ arrangement.

As with recent molecular biology studies, mathematical and physical models of phyllotaxis continue to be developed (Pennybacker *et al.*, 2015). These physical models, which are based on the mechanical forces acting at the level of the SAM, deal mainly with the close packing of elements on a surface and with the emergence of regular phyllotactic patterns where there is no stochastic variation in the phyllotactic patterns. However, recent combinatory approaches used to analyse phyllotactic patterns took into account the random fluctuations in the angle of divergence as observed in living systems (Guédon *et al.*, 2013). In other cases, semi-chaotic phyllotactic patterns observed in nature were recognised in the theoretical framework of dynamical systems (Atela *et al.*, 2002). The empirical and theoretical studies of irregular phyllotactic patterns will certainly lead to new research areas in this discipline.

Despite the development of new models and the use of new methodologies, there are few recently published books dedicated exclusively to phyllotaxis. The book *La phyllotaxie* (Loiseau, 1969) provides a survey of the experimental studies dealing with the establishment and characterisation of phyllotactic organisation. The book *Phyllotaxis* (Jean, 1994) provides an excellent account of the theoretical developments that took place during the second half of the 20th century and contains extensive bibliography. Likewise, the multi-authored book *Symmetry in plants* (Jean and Barabé, 1998a) contains several articles on the diverse aspects of phyllotaxis, covering both experimental and theoretical approaches. However, nearly all of the papers published in this book deal with regular phyllotactic patterns. Additionally, no general synthesis dealing with the actual status of phyllotactic research has been published since. Because the reader can find a comprehensive review of the literature on phyllotaxis in Loiseau (1969), Rutishauser (1981), and Jean (1994), we focus the present survey on the past 20 years, which are characterised by the discovery of new biological processes and by the elaboration of original theoretical models for phyllotaxis. During this period, two main types of theoretical models were developed: (1) those dealing mainly with the geometrical and dynamical properties of phyllotactic organisation, without direct link to the biological or physical processes involved in the emergence of phyllotactic patterns in plants, and (2) those incorporating biological and physical variables and parameters that may be involved in the processes governing the structure of the SAM. The present book focuses mainly on the second type of models developed for the phyllotaxis of plants. Although some of these models were developed more than ten years ago, we believe it is important to present them in detail. These models integrate new biological or physical principles acting at the level of the SAM and mark an important step in the history of phyllotaxis. They will likely stimulate newcomers in phyllotaxis to expand them an/or to develop new original approaches to study phyllotactic patterns.

The main goals of this book are: (1) to provide a survey on the recent developments in the study of phyllotaxis, (2) to highlight substantial advances within the discipline, and (3) to provide a comprehensive summary and insight into the discipline for the novice as well as experienced students of phyllotaxis.

The book is divided into six chapters:

Chapter 1, entitled *Phyllotactic Parameters*, outlines a comprehensive list of parameters (mainly geometrical) that have been developed and used to describe and analyse phyllotaxis. These parameters, which are essential for the description and understanding of phyllotaxis, are discussed in the context of current models.

Chapter 2, entitled *Dynamical Models*, discusses theoretical phyllotactic models that are based on energetic principles and iterative processes and that can generate regular phyllotactic patterns. For example, we discuss Douady and Couder's as well as Levitov's novel dynamical models that served as a foundation for other approaches.

Chapter 3, entitled *Statistical and Probabilistic Approaches*, explores the probabilistic features of phyllotactic models and their predictive value. We focus on the manifestation of irregular patterns that appear in several plants, including mutants, and on statistical methods used to analyse the degree of order of these patterns.

Chapter 4, entitled *Role of Genes in the Framework of Biochemical and Molecular Models*, outlines the recent molecular findings that might explain the processes involved in pattern initiation, development, and establishment at the level of the SAM. We focus on theoretical models based on molecular data involving auxin and it's feedback mechanisms, as well as the genes implicated in phyllotactic patterning.

Chapter 5, entitled *Biophysical Aspects of Phyllotaxis*, discusses the biophysical processes involved in the establishment of phyllotactic patterns. In particular, we discuss recently developed biomechanical models in detail; these models are based on physical laws linked to compressive and tensive forces on a metallic plate. We also discuss the analogy between the appearance of primordia on a SAM and buckling on a plate and present joint models involving both biomechanical and biochemical processes.

Chapter 6, entitled *Concluding Remarks: Critical Analysis and State of the Discipline*, summarises the interrelationships between the various aspects and approaches discussed in this book. It also highlights unsolved problems in this field, such as the wide variety of SAM morphologies and associated unexplored lateral organs.

1 Phyllotactic Parameters

Organ Initiation and Functioning of the Shoot Apical Meristem

The shoot apical meristem (SAM) is the main growth centre that produces appendicular structures or lateral organs such as leaves in very specific and precise arrangements. It is often referred to as a system with indeterminate growth that generates lateral organs with determinate growth. The diversity of phyllotactic patterns is limited by a number of genetic and physical constraints. These issues and their impact on the manifestation of phyllotactic patterns are discussed in detail in all chapters.

The SAM of angiosperms is typically divided into the central and peripheral zones (Bowman and Eshed, 2000; Steeves and Sawhney, 2017). The central zone (and underlying rib zone) is characterised by a group of slowly dividing cells that act mainly as reserve cells. The peripheral zone consists of actively dividing cells, corresponding to the zone where lateral organs such as leaves are initiated. Superimposed on this general zonation pattern are the characteristic cellular arrangements referred to as the tunica and corpus. The tunica typically consists of two or three layers of cells that divide in a plane perpendicular to the surface of the meristem (technically described as anticlinal). The corpus represents the underlying cells that divide in a variety of planes (Steeves and Sussex, 1989; Steeves and Sawhney, 2017). These zones are

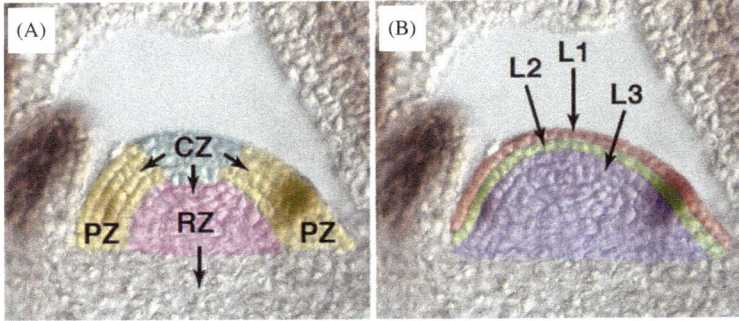

Fig. 1.1. Histology of the shoot apical meristem (SAM). **(A)** Lateral organs are produced from cells recruited from the peripheral zone (PZ), whereas the bulk of the stem is derived from cells recruited from the rib zone (RZ; the outermost layers of the stem are derived from the peripheral zone). The central zone (CZ) acts as a reservoir of stem cells that replenish the PZ and the RZ and maintain the integrity of the CZ itself. **(B)** The SAM is composed of clonally distinct layers of cells. In the SAM of eudicots, there are typically three layers whereas that of many monocots, including grasses, is composed of only two layers. The integrity of the epidermal layer (L1) and sub-epidermal layer (L2), which form two clonally distinct layers, is maintained by the almost exclusive anticlinal cell division orientation within each layer. The L1 and L2 are collectively referred to as the tunica. Cells interior to the L2 constitute the corpus (L3), in which various orientations of cell division are observed. *Filamentous Flower (FIL)* expression (brown colour) evidences lateral organ anlagen in the PZ and the abaxial domains of leaf primordia in Arabidopsis. (From Bowman and Eshed, 2000; reprinted with permission from Elsevier.)

well-characterised anatomically (Fig. 1.1) and have recently been characterised genetically (Wang and Li, 2008; Traas, 2019).

The action of specific genes implicated in the generation of phyllotactic patterns has been the object of much research in recent years (see the chapter on the "Role of Genes in the Framework of Biochemical and Molecular Models"). Their activities and domains of expression in the specific zones of the SAM are described in Fig. 1.2. The interplay among these genes and other structural features of the shoot system results in the diverse

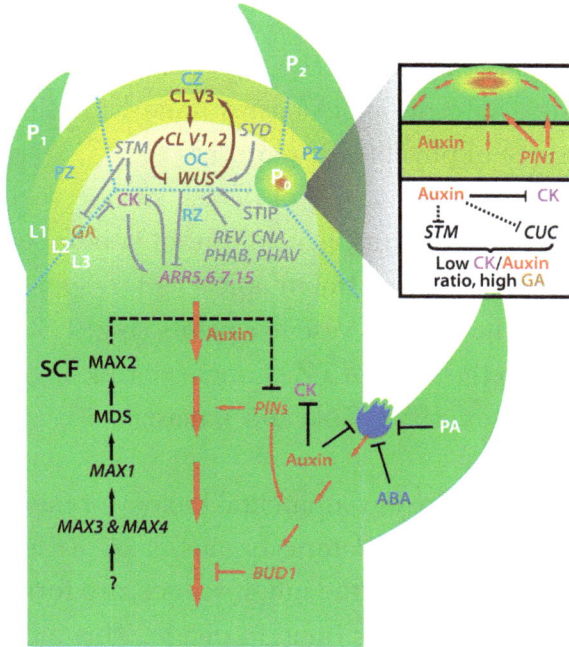

Fig. 1.2.   Pathways controlling shoot apical meristem (SAM) activities and the initiation and outgrowth of axillary meristems (AMs) and their respective elements. Pathways and elements controlling SAM activities and the initiation and outgrowth of AMs in Arabidopsis. L1–L3 cell layers and developmental zones are also shown. Brown arrows indicate the CLAVATA-WUSCHEL (CLV-WUS) feedback loop. Red arrows indicate auxin flow and its related pathway. Black arrows represent the more axillary branching (MAX)-dependent signal (MDS) pathway. Pink characters represent cytokinin (CK)-related regulators. The blue region denotes the axillary bud.

ABA, abscisic acid; ARR, Arabidopsis response regulator; BUD1, BUSHY AND DWARF1; CLV1–3, CLAVATA1–3; CNA, CORONA; CUC, CUP-SHAPED COTYLEDON; CZ, central zone; D3, DWARF3; D10, DWARF10; GA, gibberellin; HTD1, HIGH-TILLERING DWARF1; LA1, LAZY1; MOC1, MONOCULM1; OC, organising centre; OSH1, *Oryza sativa* HOMEOBOX1; OsTB1, *O. sativa* TEOSINTE BRANCHED1; P0–P5, leaf primordial; PIN, PIN-FORMED; PZ, peripheral zone; PHAB, PHABULOSA; PHAV, PHAVO-LUTA; SCF, Skp1-Cul1/Cdc53-F-box; STM, SHOOT MERISTEMLESS; SYD, SPLAYED; TA, tiller angle; TAC1, TILLER ANGLE CONTROL1; PA, polyamines; REV, REVOLUTA; RZ, rib zone; WUS, WUSCHEL. (From Wang and Li, 2008; reprinted with permission from Annual Review Inc.)

predictable and stable patterns of phyllotaxis. The study of the genetic mutations that affect the expression of these genes has provided insight into the mechanisms underlying the generation and maintenance of phyllotactic patterns.

The SAMs of other groups of plants, such as seedless vascular plants and gymnosperms, have different general organisations that range from a single central or apical cell (responsible for pattern generation) in seedless vascular plants (Fig. 1.3C) to a structure analogous to the CZ and PZ of angiosperms in gymnosperms (Steeves and Sussex, 1989; Gola and Banasiak, 2016; Steeves and Sawhney, 2017).

One feature of the developmental biology of meristems characteristic of plants is that lateral organs, such as leaves on the shoot or floral primordia on the inflorescence, are formed and 'left behind' while SAM cells continue to divide. This mode of development is analogous to the stacking of building blocks (i.e. cells with cell walls). Leaf primordia (which eventually grow into determinate structures) are characterised as a group of cells that divide and differentiate further into a 'unit' or lateral organ associated with a portion of the shoot. This portion of the shoot, also known as the phytomer or node-internode, is 'left behind' as the meristem (with indeterminate growth) continues to divide and produce new units. Thus, the movement or flow of cells is not a feature of plants.

In addition to vegetative shoot systems, phyllotactic patterns have also been documented in inflorescences (Figs. 1.4A–B). As in shoots, specific and predictable phyllotactic patterns emerge in inflorescences and characterise particular taxa. Our work on the inflorescences of Araceae has revealed a lattice-like arrangement of floral buds forming distinct intersecting spirals of structures along the entire length of the meristem (Fig. 1.4A). The inflorescences of Asteraceae have also been studied from a descriptive mathematics perspective. Although

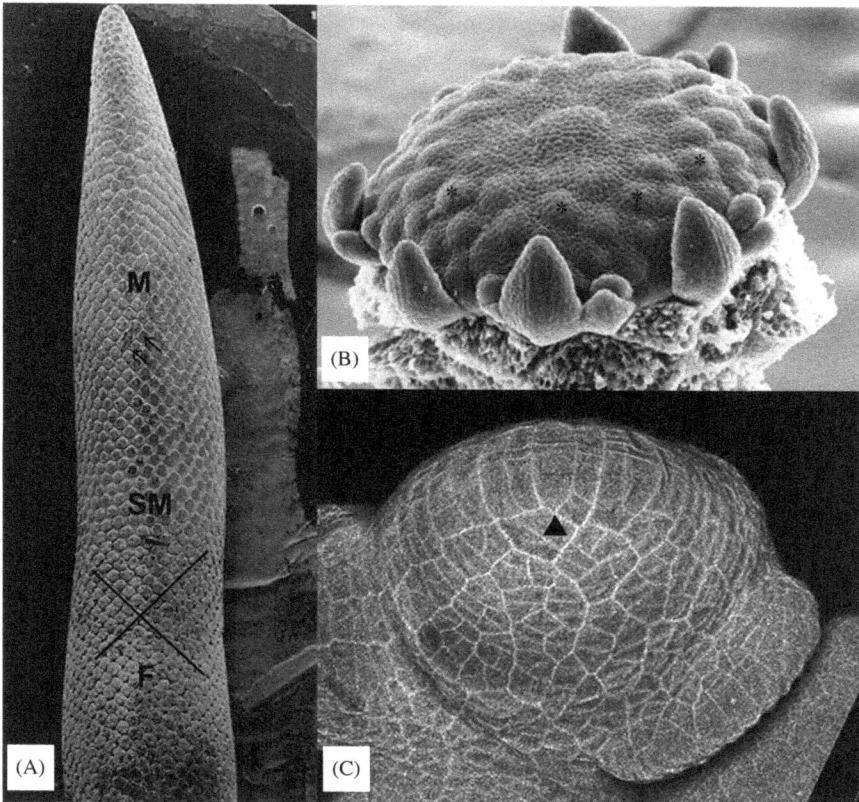

Fig. 1.3. Variations of the meristematic structures producing phyllotactic patterns. **(A)** Cylindrical inflorescence of *Philodendron ornatum* showing a lattice of floral meristems representing male (M), sterile male (SM), and female (F) floral zones arranged in parastichies (black lines). Interruptions in these parastichies are visible along the inflorescence as the diameter of the structure varies (arrows). Magnification, 35×. (From Barabé and Lacroix, 2008; reprinted with permission from University of Chicago Press Journals.) **(B)** Side view of *Symphyotrichum laurentianum* inflorescence. Floral buds (asterisks) are initiated in a less regular pattern and parastichies are not obvious. Magnification, 630×. (From Lacroix *et al.,* 2007; reprinted with permission from Canadian Science Publishing.) **(C)** Side view of the SAM of *Selaginella kraussiana* showing the distinct meristematic cells (arrowhead) typical of many seedless vascular plants. Magnification, 375×. (From Jones and Drinnan, 2009; reprinted with permission from University of Chicago Press Journals.)

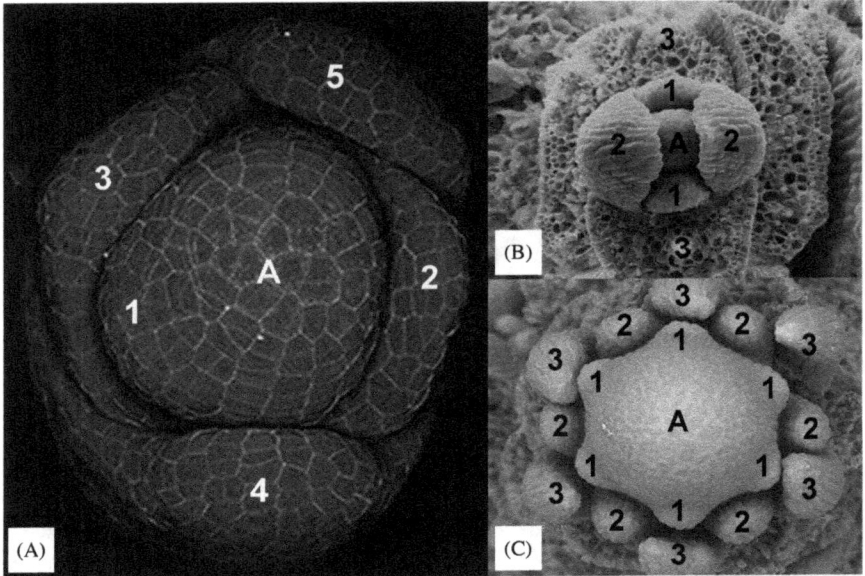

Fig. 1.4. Common phyllotactic patterns at the level of the shoot apical meristem (SAM) in angiosperms. The numbers on individual leaf primordia indicate the relative age of the leaf, with the lowest number corresponding to the youngest structure. (**A**) Top view of *Aponogeton madagascariensis* SAM (A) showing the spiral pattern. Magnification, 200×. Photo courtesy of A. Dauphinée, Dalhousie University. (**B**) Top view of *Thuja occidentalis* SAM showing the decussate leaf arrangement, in which leaves are initiated in successive pairs. Magnification, 200×. (From Lacroix *et al.*, 2004; reprinted with permission from Canadian Science Publishing.) (**C**) Top view of *Myriophyllum aquaticum* SAM showing three whorls of six leaves. Magnification, 310×. (Photo courtesy of C. Lacroix, University of Prince Edward Island.)

the concave form of the developing inflorescences of *Helianthus* sp. (sunflower) is well known, the young inflorescences of other genera (e.g. *Aster*; Fig. 1.4B) are characterised by a convex dome. Ovulate (female) pine cone morphology in gymnosperms has also been the object of mathematical analysis. The 'universality' of specific patterns of organ arrangement on the meristem of several land plants (Gola and Banasiak, 2016) has led to the development of a notation system that has been used to describe, analyse, and model phyllotaxis.

Lateral organs such as leaves are initiated as more or less centric bumps on the periphery of the SAM (Fig. 1.4). The location of these primordia follows a very specific pattern (i.e. spiral, whorled) and it can be predicted based on the position of previously formed primordia. Depending on the relative size of the SAM and primordia, the extent of structure packing on the shoot system varies greatly. For example, in *Begonia* (Fig. 1.5A)

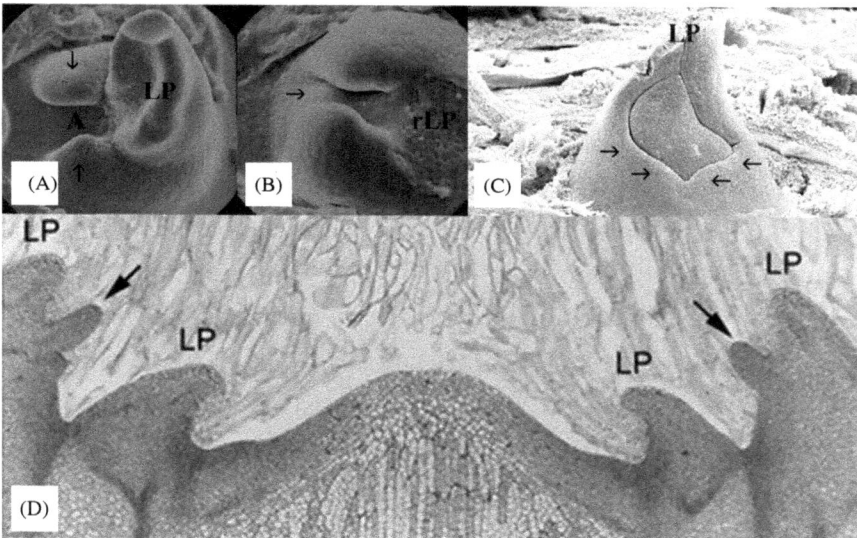

Fig. 1.5. Variations of leaf shape and size in relation to the shoot apical meristem (SAM) in angiosperms. (**A**) Top view of the SAM of *Begonia scabrida* (distichous phyllotaxis) showing a leaf primordium (LP) with a broad leaf base (arrows) in relation to a relatively small SAM (A). Magnification, 140×. (From Barabé *et al.*, 2007; reprinted with permission from Oxford University Press.) (**B**) The later developmental stage of the same system showing the initiation of a new primordium (arrow) on the relatively small SAM. rLP, removed LP. Magnification, 180×. (From Barabé *et al.*, 2007). (**C**) Side view of the SAM of *Euterpe oleracea* (spiral phyllotaxis) showing an encircling leaf base (arrows) typical of many dicotyledonous plants. Magnification, 100×. (From Barabé *et al.*, 2010; reprinted with permission from Canadian Science Publishing.) (**D**) Longitudinal section of the SAM of cactus *Oroya depressa* showing a relatively large meristem flanked by LP with initiating spines (arrows). (Magnification, 65×. Photo by James Mauseth (http://www.sbs.utexas.edu/mauseth/weblab/webchap6apmer/6.3-7a.htm — accessed July 2016.)

and *Euterpe* (Fig. 1.5B) species, the base of individual leaf primordia can take up a significant portion of the meristem at the time of primordia initiation and early development. In other systems, such as those that form whorls of multiple leaves, the relatively large meristem maintains a visible, characteristic form throughout the cycle of initiation of consecutive primordia (Fig. 1.3C).

Two main factors influence the position of lateral organs: the position of the other organs on the apex, and the shape of the apex itself. These factors were analysed in detail in developmental studies of the SAM of aquatic plants (Kelly and Cooke, 2003) and of flowers of different families of monocots (Kirchoff, 2000, 2003), and documented in a comprehensive review of the diversity of the phyllotactic patterns in land plants (Gola and Banasiak, 2016). Two general rules for the initiation of new primordia have been proposed by botanists. The first, Hofmeister's (1868) rule, proposed that each new primordium appears in the area furthest away from the base of existing primordia. The second rule, formulated by Snow and Snow (1931, 1962), stipulates that each new primordium, covering a minimal area, appears between two older primordia where there is enough space for its initiation. This rule is well-known as the theory of the first available space. Both rules have been used consistently in the construction of theoretical dynamic models of phyllotaxis (e.g. Douady and Couder, 1992, 1996a–c) but they may not apply to all existing phyllotactic patterns. For example, on cacti showing 'ribbed spiral patterns' and in mutants showing random patterns, these rules might not hold. Douady (1998) analysed the mathematical relationship between the two rules with regard to the appearance of particular phyllotactic patterns.

Leaf Asymmetry and Phyllotaxis

The foliar primordia of many plants are asymmetrical, a feature that appears to be particularly common in monocots (Hirmer, 1922). For example, Korn (2006) reported that some Araceae (i.e. *Syngonium podophyllum*, *Aglaonema crispum*, *Dieffenbachia maculata*, *Dieffenbachia amoena*, and *Epipremnum aureum*) have asymmetrically sized lamina. This phenomenon has also been observed in the Maranthaceae *Calathea ornata* (Korn, 2006) and in the Arecaceae *Euterpe oleracea* (Barabé *et al.*, 2010), among others. In all these plants, the larger side of the younger leaf faces the smaller side of the older leaf, independently of the direction of the genetic spiral. In plants with asymmetrical leaves, the smaller half lamina on the anodic side of the leaf blade indicates that the asymmetry is of phyllotactic origin (Dormer, 1972; Korn, 2006).

There are important developmental interactions between the asymmetry of leaf primordia and the functioning of the SAM. According to Korn (2006), the presence of slightly asymmetrical young leaves would be due to the delayed growth of the anodic side of the leaf blade, which might, in turn, be explained by the presence of a more robust vascular trace on the cathodic side. Tamaoki *et al.* (1999) have shown that the over-expression of the *Nicotiana tabacum homeobox 1* gene (*nth1*) in transgenic tobacco results in abnormal leaf morphology, characterised by the wrinkled and curved lamina. There is a correlation between the direction of the generative spiral and the direction of the curvature in the asymmetric leaves of tobacco (Tamaoki *et al.*, 1999). In mutants with a clockwise generative spiral, malformed leaves are curved to the right, whereas in mutants with a counterclockwise spiral, leaves are curved to the left (Tamaoki *et al.*, 1999). These researchers were also able to show that SAM activity is essential for leaf asymmetry. In rice, the mutant *leaf lateral asymmetry 1* is characterised by leaves with a reduced half lamina on the side facing the next younger primordium (anodic side), and the asymmetric growth of leaf primordia is due to an aberrant SAM organisation (Obara *et al.*, 2004).

The role of auxin in a phyllotactic organisation (Kuhlemeier, 2007) and the formation of vascular bundles (Bayer *et al.*, 2009) have been clearly established. Therefore, it is not surprising that leaf asymmetry can also be explained by processes involving auxin. Chitwood *et al.* (2012) showed that leaf asymmetry in *Solanum* and *Arabidopsis* species is due to developmental constraints imposed on leaf morphology by auxin-dependent phyllotactic patterning, resulting from an unequal distribution of auxin on both sides of the primordium. According to these authors, the first primordium (P1) is affected by the next auxin maximum, which will give rise to another primordium (P0). As the size of P1 increases, its sink effect is reduced, and the P1 auxin peak is shifted away from the new auxin peak (P0) (Fig. 1.6). Therefore, as the concentration of auxin diminishes on the side of P1 facing the emergent primordium (P0), there is an increase on the descending side of P0, which leads to the bilateral asymmetry of the leaf. This is in agreement with the direction of asymmetry predicted by the theoretical model and obtained from quantitative measurements in tomato (*Solanum lycopersicum*), where the higher auxin concentration is found on the descending side of the primordium. This is a consequence of the mode of action of auxin in phyllotactic patterning, which constrains laminar form and growth on both sides of the leaf. The direction of the phyllotactic spiral determines the side of the leaf where the asymmetry will occur. The interpretation of Chitwood *et al.* (2012) could probably be generalised to all spiral phyllotactic systems.

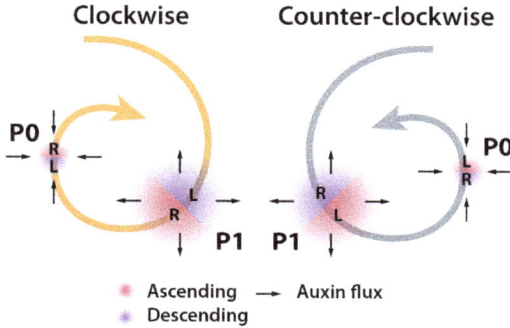

Fig. 1.6. The left and right sides of leaf primordia are exposed to differing auxin concentrations, depending on phyllotactic direction. Diagram of clockwise (orange) and counterclockwise (grey) phyllotactic systems. Ascending (red) and descending (purple) sides, as well as left (L) and right (R) sides of P0 and P1 are denoted. Black arrows indicate auxin flux. The relationship between young and old leaf primordia is such that auxin is depleted from the ascending side of older primordia and supplied to the descending side of younger primordia. (From Chitwood *et al.*, 2012; reprinted with permission from American Society of Plant Physiologists.)

Note: (1) Different designations have been used to describe the direction of the genetic spiral. Chitwood *et al.* (2012) defined the direction by ascending the spiral (clockwise or counterclockwise) from older to younger leaves. Along the spiral, they named the side of the leaf facing younger leaves 'ascending' and the side facing older leaves 'descending', which correspond to the anodic side (+) and cathodic (–) sides of Snow (1965) and Korn (2006), respectively.

Although the distribution of auxin on both sides of the primordium depends on the direction of the genetic spiral (Fig. 1.6), the relationship between phyllotaxis and leaf asymmetry is certainly affected by other phyllotactic parameters, particularly when leaf asymmetry is highly accentuated, as is the case in *Begonia* species (Barabé *et al.*, 2007). In nature, typical asymmetrical leaves are often associated with a morphological phenomenon known as 'Pendelsymmetrie' or pendulum symmetry. This term, coined by Goebel (1928), designates developmental patterns with some kind of left/right oscillation in symmetry (Charlton, 1998). The concept of pendulum symmetry also includes alternate positioning of asymmetrical leaves in spiro-distichous phyllotactic patterns, such as those observed in *Ulmus* (Charlton, 1993) and *Begonia* (Barabé *et al.*, 1992, 2007) species, where leaf initiation oscillates between two sectors of the SAM. Pendulum symmetry was studied in detail by Charlton (1998) from both ontogenetic and morphological perspectives. However, the developmental processes linking the asymmetry of the leaf and the dorsiventrality of the shoot have yet to be identified. The particular phyllotactic patterns (distichous vs. spiral), the ratio between the size of the SAM and primordium [referred to as parameter b by Van Iterson (1907)], and the plastochrone are morphological parameters that are most likely linked to asymmetry. For example, Tamaoki *et al.* (1999) found a positive correlation between the degree of leaf curvature and the value of the plastochrone ratio in the tobacco mutant *nth1*. However, the relationship between biochemical processes governing leaf symmetry and phyllotactic patterns remains an open question.

Fluctuating Form and Size of the SAM: The Case of *Thuja Occidentalis*

According to Green and Baxter (1987), during a cycle of leaf production, the surface of the shoot apex enlarges to a maximum and a 'crease' forms, demarcating the peripheral location of the LP. This developmental event results in a reduction of the surface of the apex to a minimum. The repetitive cycle of initiation of primordia and size variation of the shoot apex is common across a variety of phyllotactic patterns. This phenomenon is clearly visible in the SAM of *Thuja* where two types of leaves are alternately produced (Lacroix *et al.*, 2004).

In *Thuja occidentalis*, the SAM is bilateral in symmetry and bears leaves in an orthogonal decussate phyllotactic pattern. Two alternating and morphologically different pairs of leaves constitute the basic repeating pattern of the SAM (Fig. 1.4B). At maturity, the dimorphism between leaf types is marked. Leaves in one plane are wide and flat contrasting with the narrow and cup-shaped leaves growing in the perpendicular plane. During the earliest visible stages of initiation (primordial crest), cup-shaped and flat leaves are very similar in morphology. As individual leaf primordia become more delimited, the two leaf types (flat vs. cup-shaped) become morphologically distinct. Analysis of SAM diameter, angle of insertion of individual leaves, and size of leaf primordia (in both the tangential and perpendicular planes) showed that, during the early stages of development, flat and cup-shaped leaves are morphologically similar, diverging in their pattern of development post-initiation, especially in terms of leaf width and thickness. In *Thuja* (Lacroix *et al.*, 2004), the size and shape of the apex go through a 'repeating' cycle that is related to the type of primordium initiated. This shows that the size and shape of the apex can change on the same plant depending on the type of leaves produced. This cycle, therefore, accounts for the production of two types of leaves and covers two plastochrones.

Description of Spiral Phyllotactic Patterns

The presence of particular sequences of numbers and angles between successive primordia in spiral phyllotactic patterns has long attracted research attention in many disciplines (see the historical section on "Introduction"). At the beginning of the 19th century, Schimper (1830) and Braun (1831) proposed an ideal cylindrical phyllotaxis representation. In this representation, leaves

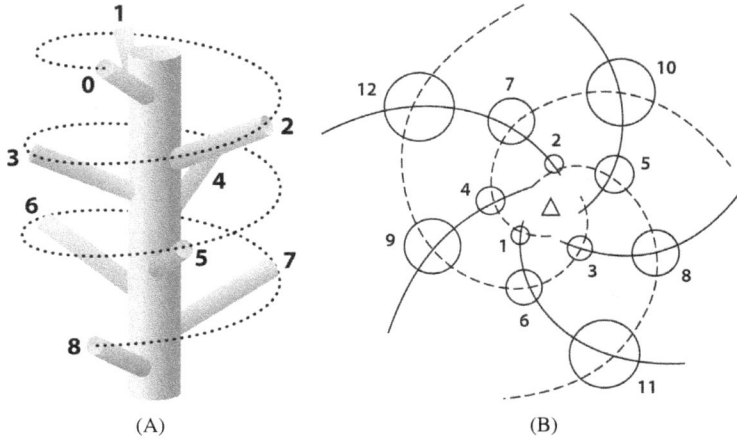

Fig. 1.7. (**A**) Cylindrical representation of a phyllotactic system (3, 5) with a phyllotactic fraction of 3/8. The dotted line represents the generative spiral. (**B**) Planar diagram of a shoot system with a (3, 5) spiral phyllotactic pattern. There are three clockwise parastichies (dotted lines) and five counterclockwise parastichies (solid lines).

are placed along a genetic spiral in their order of appearance and are separated by a roughly constant divergence angle. Phyllotaxis is characterised by a rational fraction linked to the divergence angle. For example, in Fig. 1.7A, the phyllotactic fraction of the phyllotactic pattern is 3/8. In this example, leaves 0 and 8 are superposed in a vertical row; this is called orthostichy. There are eight leaves from leaves 0 to 8 (not counting the first leaf), and the genetic spiral winds around the stem three times. In this example, each 'cycle' consists of eight leaves and three turns. If there are enough leaves around the stem, eight orthostichies (vertical rows) will be visible. In the case of a 1/2 pattern, which corresponds to a distichous system (two orthostichies), each cycle comprises two leaves and one turn. The most frequent fractions observed in nature belong to the main sequence 1/2, 1/3, 2/5, 3/8, 5/13, ..., where the numerators and denominators are numbers of the Fibonacci sequence

(1, 1, 2, 3, 5, 8, …). In this series, each number is the sum of the two preceding ones. Spirals joining leaves are another way of looking at the positional relationships between primordia at different stages of development of a shoot system. Two sets of spirals, called parastichies, run in opposite directions, and each lateral organ is located at the intersection of two parastichies (Fig. 1.7B).

A spiral phyllotactic pattern can be described by the number of parastichies found in each set. For example, Fig. 1.7B shows three parastichies winding in one direction (dotted lines; linking leaves 3, 6, 9, and 12; 2, 5, 8, and 11; and 1, 4, 7, and 10) and five parastichies winding in the other direction (solid lines). This system is referred to as a (3, 5) phyllotactic pattern. These numbers are (typically) consecutive numbers in the Fibonacci series (1, 1, 2, 3, 5, 8, …), highlighting the regularity of leaf initiation sites and their predictable position in relation to one another. The phyllotactic fraction can be calculated using a planar representation of this pattern. In this case, the numerator of the fraction represents the number of times one has to circle around the stem to move from one leaf to the one directly above it. As shown in Fig. 1.7A, we obtain the fraction 3/8. Multiplying this fraction by 360 gives an average divergence angle (i.e. the horizontal angle between two successive leaves) of 137.5° (Fig. 1.8A). The series of Fibonacci fractions (1/2, 1/3, 2/5, 3/8, 5/13, …) tends to a divergence angle of 137.5°, also called the golden angle. The Fibonacci sequence (1, 1, 2, 3, 5, 8, …), golden angle (137.5°), and opposed parastichies (m, n) linking primordia in a shoot system have been noted to occur consistently and persistently in a wide variety of plant species (Gola and Banasiak, 2016; Peaucelle and Couder, 2016). Other phyllotactic patterns such as whorled and distichous systems can also be described using similar parameters, as discussed in the next section.

Phyllotactic Parameters

Three general types of regular phyllotactic patterns are commonly seen in plants: spiral (leaves inserted in a spiral around the stem), whorled (two or more leaves inserted at the same node), and distichous (leaves inserted alternately in two rows on opposite sides of the stem). Variations on these main patterns, such as spirodistichy (distichous system forming a winding spiral along the stem) or bijugy (double Fibonacci systems), are also seen. Disorganised phyllotactic patterns that occur naturally (Endress, 1989; Rutishauser, 2016) or that are produced artificially in mutants (Itoh *et al.*, 2000) must also be considered. Further, botanical classifications of phyllotaxis do not necessarily correspond exactly to theoretical mathematical models. For example, in Jean's mathematical model (1994), distichous patterns are considered a type of spiral pattern and whorled patterns are characterised as transitional unstable systems. Although there is a continuous transition from distichous patterns to spiral patterns in nearly all mathematical models, we consider them morphologically distinct. In other words, morphologically distinct categories do not necessarily correspond to how these patterns are related mathematically.

The use of specific parameters to quantitatively analyse phyllotactic patterns provides a common framework on which researchers can base their models and hypotheses (Jean, 1994; Rutishauser, 1998). Four basic parameters are highlighted here (Fig. 1.8): (1) divergence angle, (2) contact parastichy pattern, (3) parameter b, and (4) plastochrone ratio. Geometric patterns and the parameters that characterise them are discussed in detail below.

A rigorous geometrical framework is needed for a precise biological analysis of phyllotactic patterns. Two types of representations are generally used: planar (Figs. 1.7 and 1.8) and cylindrical

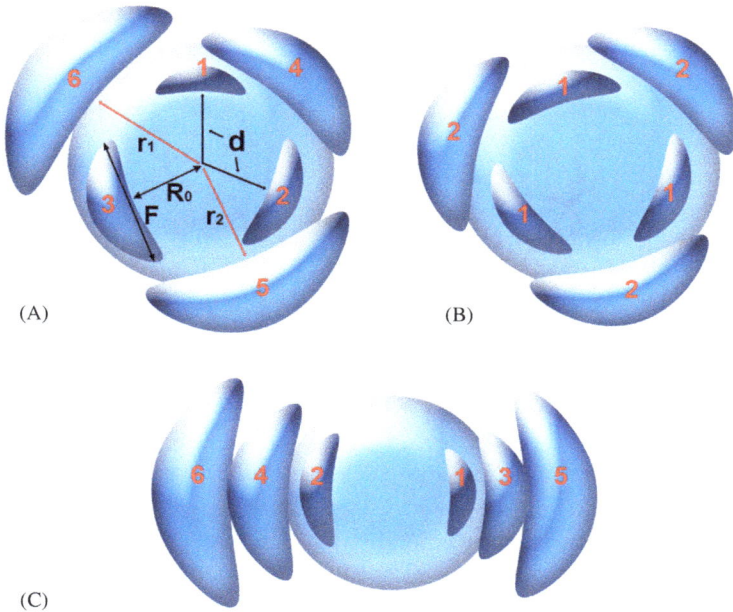

Fig. 1.8. Diagrammatic representation of phyllotactic patterns and typical phyllotactic parameters. (**A**) Spiral patterns. (**B**) Whorled phyllotactic pattern with three leaves per whorl. (**C**) Distichous pattern where each leaf is initiated at 180° from the previous one. Leaf number refers to the sequence of initiation, 1 being the youngest primordium. d is the divergence angle between leaf primordia 5 and 6. r_1 and r_2 are distances from the centre of primordia 1 and 2, respectively, that are used to calculate the plastochrone ratio (R). R_0 is the radius of the SAM. F is the primordial width used to calculate parameter $b = \frac{R_0}{F}$.

(Fig. 1.9). Jean (1994) mathematically analysed the geometrical parameters involved in these representations. Depending on the nature of the biological material used, one representation will be more adequate than another. For example, if we deal with a typical apex that is roughly conical, like that of *Arabidopsis* or *Solanum*, a planar representation is more appropriate because phyllotactic parameters can be empirically measured in a cross-sectional plane. By contrast, if lateral organs are inserted on a cylindrical structure, like the spadix of the Araceae (Fujita, 1942; Jean and Barabé, 2001) or the *Magnolia* flower (Zagórska-Marek, 1994), a cylindrical representation

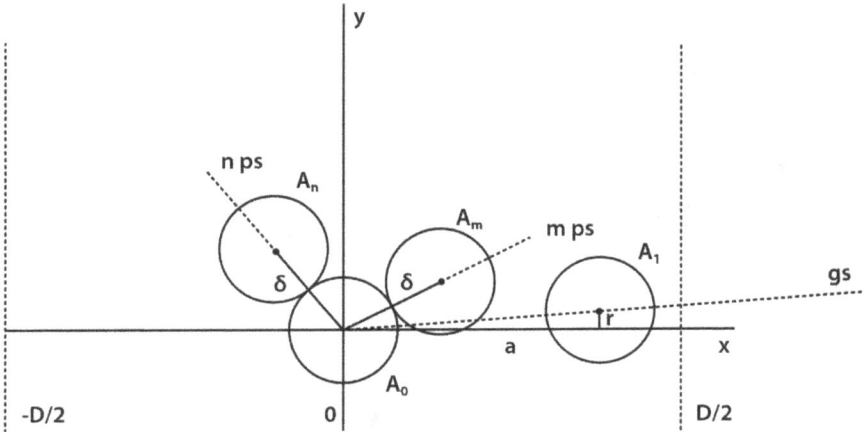

Fig. 1.9.   Cylindrical representation of phyllotaxis. The origin is placed at the centre of the band, so that the horizontal positions are between $+D/2$ and $-D/2$, where D is the circumference of the cylinder. The position of the first element on the generative spiral A_1 is given by the divergence a and the rise of the generative spiral r. The element at the origin, A_0, is in contact with two other elements, A_m and A_n, where m and n are the respective numbers of parastichies in the two directions (*m* ps and *n* ps.) (From Douady, 1998; reprinted with permission from World Scientific Publishing.)

is more appropriate. The geometric parameters are essentially the same in both representations. However, those that do not look biologically similar (e.g. plastochrone ratio, rise, and parameter b) are linked mathematically. For example, the rise r is related to the plastochrone ratio R as $\frac{\ln R}{2\pi}$ (Jean, 1994). Researchers have developed theoretical models using either type of geometrical representation depending on the specific problems to be addressed.

If the values of certain phyllotactic parameters are known, the theoretical values of other parameters can be calculated (see Table 4.2 in Jean [1994]). For example, theoretically, an increase in the divergence angle is correlated with an increase in the plastochrone ratio, rise, or parameter b, as supported by many observations in plants. In a survey of phyllotactic patterns in various species, Rutishauser (1981, 1982, 1998) used these parameters to determine whether there were any correlations between their values and the expression of specific phyllotactic patterns. This is only one example of the use of these parameters to investigate the expression of phyllotactic patterns.

Generative Spiral and Divergence Angle

The generative spiral or fundamental spiral (Schimper, 1830; Braun, 1831) is the line along which leaves are periodically produced with a constant divergence angle (d) (Fig. 1.8).

In a cylindrical representation (Fig. 1.9), the cylinder can be unrolled in a single plane. Disks of diameter δ are placed regularly (at a constant height and angle from the preceding one) along a generative spiral (a helix in the case of a cylinder). Therefore, the circumference of the cylinder (D), the diameter of the disk (δ), and rise of the generative spiral between two successive elements (r) can be measured. A generative spiral can be defined by the horizontal distance on the circumference of the cylinder (a) and the rise between two successive elements (r).

Botanists and mathematicians began quantitatively analysing phyllotaxis owing to the observed regularity of different types of organs. The frequent occurrence of a divergence angle of 137° and the fact that parastichy numbers belong to the Fibonacci series remain actively studied topics (Okabe, 2016). For example, Okabe (2011) considered that the main Fibonacci sequence is selected because it involves fewer phyllotactic structural transitions as a plant grows to maturity. He later proposed that the golden angle observed in phyllotaxis is the optimal solution to minimise the energy cost of phyllotactic transition (Okabe, 2015a; Okabe *et al.*, 2019). This explanation is similar to the minimal entropy concept in Jean's phyllotactic model (1994). By contrast, Cooke (2006) geometrically analysed the Fibonacci sequence in phyllotaxis based on empirical data from different plants and disagreed with the interpretation that Fibonacci-related patterns are linked to a global imperative for optimal packing of foliar primordia. Instead, he considered that phyllotactic patterns arise from primordia already positioned on the SAM, indicating the role of biological interactions.

Okabe (2015b) accurately measured the divergence angle between florets in *Helianthus annuus* (sunflower) capitula and found a value of $d \cong 137.513 \pm 0.003°$ and described the phyllotactic system as (34, 55). However, the phyllotactic pattern of the capitulum may not be as regular as believed. After observing a large number of *H. annuus* capitulum samples ($n = 657$), Swinton *et al.* (2016) determined that nearly 20% of parastichy numbers did not correspond to the Fibonacci series (1, 2, 3, 5, 8, 11, ...). Battjes and Prusinkiewicz (1998) geometrically analysed *Microsotis* flower heads and observed that the divergence angle shows regular non-random deviations from the average angle of 137°. However, these deviations are intrinsic properties, and no exceptions, of the capitulum's phyllotactic pattern. Further, when a large number of primordia exist on a planar surface, a very small change in the divergence angle is known to generate different phyllotactic patterns (Prusinkiewicz and Lindenmayer, 1990). In relation to the occurrence of the Fibonacci sequence in plants, Fierz (2015) observed the number of parastichies (m, n) in 6000 *Pinus nigra* cones and noted that 97% of cones followed the main Fibonacci sequence (8, 13). Further, the author found nine aberrant spiral patterns with Fibonacci-type sequences such as (7, 11) and (9, 13). The author also calculated the deviation of m/n from the golden ratio $\phi = 0.618$ and found that, with very few exceptions, the greater the deviation, the rarer is the pattern. Fierz (2015) also hypothesised that aberrant patterns may result from an unusual value of the primordia size/meristem size ratio at the beginning of cone development. These findings indicate that even very regular patterns contain more variability than believed (Douady and Golé, 2016; Szpak and Zagórska-Marek, 2011; Zagórska-Marek and Szpak, 2016).

Visible Opposed Parastichies

In a phyllotactic system, one can generally recognise families of parastichies running in opposite directions. Each lateral element around the centre of the apex is located at the intersection of two or more parastichies. m and n represent the number of opposed parastichies joining each element with its neighbours. A pair of families of parastichies (m, n) is called a parastichy pair (Jean, 1994). For example, in *H. annuus* capitula, opposed parastichy pairs such as (13, 21) or (34, 55) are typically seen depending on the capitulum size, with a divergence angle of 137.5° (Fig. I.1C). For the apex of *Sempervivum clacaratum*, two visible opposed parastichy pairs, (5, 3) (Fig. 1.10A) and (5, 8) (Fig. 1.10B), are seen.

When primordia are represented by points, the parastichies they form are called visible opposed parastichies because the primordia are not necessarily touching each other but are found at the intersection of any two opposed parastichies. Different visible opposed parastichies can be mapped out simultaneously in one system. For example, in Fig. 1.10, parastichies (5, 3), (5, 8), and (8, 13 — not drawn) can be represented. Among visible opposed parastichies, Adler (1974) and Jean (1994) distinguished contact parastichies (Fig. 1.10A) and conspicuous parastichies

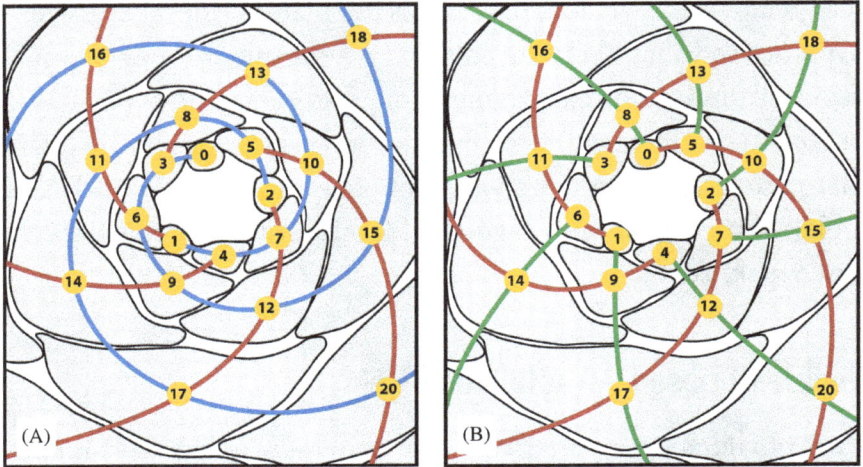

Fig. 1.10.   Transverse section of the apex of *Sempervivum calacaratum*. **(A)** Phyllotactic system with a $(5, 3)$ contact parastichy pair. **(B)** The same phyllotactic system with a $(5, 8)$ conspicuous parastichy pair. (Modified from Church, 1920.)

(Fig. 1.10B). Contact parastichies are represented by spirals joining the contact points of contiguous leaves. Conspicuous parastichies are represented by the visible parastichy pair whose angle of intersection is near 90° (Jean, 1986, 1994). This unique conspicuous parastichy pair is designated as a (m, n) phyllotactic system. The number of contact parastichies is often different from the number of conspicuous parastichies, particularly when the appendages seen in transverse sections have elongated forms (Fig. 1.10).

The numbers (m, n) can also belong to other mathematical series associated with a particular divergence angle. For example, the phyllotactic system $(5, 7)$ corresponds to the series $(2, 5, 7, 12, \ldots)$ associated with a divergence angle of 151° (Jean, 1994). The theoretical relationship between the divergence angle d and a number of parastichies is central to the Fundamental Theorem of Phyllotaxis (FTOP) developed by Adler (1974) and Jean

(1988, 1994) and recently revisited by Swinton (2012) to include all phyllotactic systems. This theorem describes the relationship between the number of conspicuous parastichies and the divergence angle between vertically successive organ positions on the lattice. According to this theorem, the divergence angle should fluctuate between limit values determined by the number of conspicuous parastichies. For example, in a (3, 5) phyllotactic system, it can vary between 120° and 144°. Theoretically, the number of parastichies will then constrain the manifestation of only a given number of divergence angles. Although this relationship holds theoretically, it does not necessarily hold empirically (Yin *et al.*, 2011). In fact, this highlights the problem of the relationship between the number of conspicuous parastichies and the number of contact parastichies from an empirical viewpoint.

Adler (1974, 1977) and Jean (1988) based their classifications of phyllotactic patterns on the notion of visible opposed pairs of parastichies that are conspicuous from a strictly mathematical viewpoint. Reick (2002) noted that Jean's (1988) and Adler's (1974) notions of conspicuous parastichies are different and result in two distinct classifications of phyllotactic patterns. However, these differences are only theoretically relevant. Reick (2002) could not determine the more appropriate notion in a botanical context. However, he noted that certain phyllotaxis models based on contact parastichies are more consistent with Jean's classification.

In theoretical geometrical models, phyllotactic patterns are characterised by conspicuous parastichy pairs, which are represented by dots on a lattice. However, in empirical or experimental studies, the phyllotactic systems are described using the number of contact parastichies, which are represented by dots on a lattice. This highlights the major difference between the botanical and the mathematical interpretations. Green and Baxter (1987) already discussed this point and noted that leaves are represented by dots in mathematical models. However, the form of leaves certainly plays a role in the dynamics of the apical meristem and consequently in the mode of formation of phyllotactic patterns, as noted in biomechanical models (see the chapter on "Biophysical Aspects of Phyllotaxis").

Parameter *b* (Van Iterson's Parameter)

Van Iterson (1907, p. 81) used parameter b ($= \delta_0/2\pi L_0$) as a measure in the geometrical analysis of pillings, where δ_0 (F in Fig. 1.8) represents the diameter of a primordium and L_0 (R_0 in Fig. 1.8), the radius of the apex. Instead of using a measure of the circumference, the diameter or radius of the apex as represented by $\Gamma = \delta_0/L_0$ by Douady and Couder (1998) can simply be used. Yin and Meicenheimer (2017) distinguished between these two parameters by calling them the circumferential (b) and radial ratio (Γ), respectively.

It is also important to consider the overall shape of lateral organs of the SAM when quantitatively analysing phyllotaxis. In a planar projection, primordia have a folioid shape, as is generally the case when viewing cross-sections of the SAM. To consider this shape in the definition of parameter Γ, Douady and Couder (1998; see also Richards, 1951) used the ratio of the azimuthal extension of the primordium (l_1) at the time of its initiation over the radius of the apex ($L_0 = R_0$ in Fig. 1.7), where $\Gamma = l_1/L_0 = \delta_0/N^{1/2} L_0$. N, which determines the conicity of the stem, corresponds to $N = \sin(\Psi/2)$, where Ψ is the angle of the cone. For $1/N = 1$, the primordium has a circular shape when projected onto a plane (see the chapter on "Dynamical Models").

The leaf arc or leaf insertion angle (*i*), a parameter comparable but not equivalent to the parameter *b*, has been used in some empirical studies to describe phyllotactic patterns (Snow and Snow, 1962; Loiseau, 1969; Rutishauser, 1998). This parameter corresponds to the angle covered by the leaf base in relation to the centre of the SAM.

Plastochrone Ratio

The plastochrone ratio ($R = r_2/r_1$ in Fig. 1.7) developed by Richards (1951) is also used commonly. It is the ratio of the distance between two successive primordia from the centre of the apex. Essentially, it is a measure of the relative growth of primordia in relation to their distance from the centre of the apex. This parameter is used extensively in empirical studies with planar representations.

In cylindrical representations, the rise is equivalent to the plastochrone ratio. Although it is not used often in empirical studies, it can be measured relatively easily in mature plants (Bursill and Rouse, 1998) and the SAM (Schwabe, 1998). As with R and b, the rise decreases with the divergence angle. In their study of the inflorescence of *Symplocarpus*, Barabé and Jean (1996) determined a theoretical correlation between R and the number of parastichies (m, n) during development.

R, b or r provide an indirect estimate of the plastochrone, which is a real-time measure separating the initiation of two successive primordia (Douady and Couder, 1992). In theoretical phyllotaxis models, b and R are often called control parameters because their continuous variation can lead to a change in the number of parastichies and the divergence angle. This phenomenon is well represented in phyllotactic diagrams consisting of sequences of bifurcations representing changing parameters (e.g. van Iterson, 1907; Adler, 1974; Douady and Couder, 1996a–c). These parameters can be varied to obtain different phyllotactic patterns ranging from distichous to spiral.

Assessing phyllotactic parameters in real-time (e.g. growth rate) remains one of the main challenges in the analysis of phyllotaxis. Barabé *et al.* (2007, 2009) analysed this problem in a study on the real-time development of *Begonia* leaves by using Green and Linstead's casting method (1990). They compared the spirodistichous pattern of *Begonia* to the spiral system of *Anagallis* and obtained a more precise estimate of the actual time involved in the formation of lateral elements. They observed that the plastochrone of *Begonia* (15 days $< T <$ 20 days) was greater than that of *Anagallis* (1.7 days $< T <$ 1.0 days); this agrees with the b and R values predicted by theoretical models.

Phyllotactic Pattern Transitions

Continuous and discontinuous transitions are two types of naturally occurring phyllotactic transitions. Continuous transitions result from symmetrical expansions or contractions of the shoot perimeter that cause uniform changes in the quantitative relationships between the primordia and the SAM (Meicenheimer and Zagórska-Marek, 1989). For example, *Linum usitatissimum* showed pattern transitions from a (5, 8) spiral pattern to a (8, 13) spiral pattern. Similarly, *Picea abies* showed a continuous transition from a (8, 13) spiral pattern to a (13, 21) spiral pattern (Rutishauser, 1998). In a continuous transition, the chirality of the genetic spiral remains unchanged (Meicenheimer, 1998). However, this transition can reduce the plastochrone ratio R and leaf insertion angle *i* (Rutishauser, 1998). Continuous transitions, whether occurring naturally, induced experimentally (Schwabe, 1971; Maksymowych and Erickson, 1977), or deduced mathematically (Adler, 1974; Jean, 1994; Pennybacker *et al.*, 2015), lead to changes in phyllotaxis from (m, n) to $(n, m + n)$ or from (m, n) to $(n - m, m)$, where $(m < n)$.

Discontinuous transitions also occur naturally (Zagórska-Marek, 1985, 1987, 1994; Meicenheimer, 1987, 1998; Jean and Barabé, 2001; Mauseth, 2004) and can be induced surgically (Snow and Snow, 1935; Loiseau, 1969) and chemically (Snow and Snow, 1937; Meicenheimer, 1981). They occur via asymmetrical expansions or contractions in localised sectors of the shoot system and result in the addition or loss of one or more parastichies in a family (Meicenheimer and Zagórska-Marek, 1989; Jean and Barabé, 2001; Szpak and Zagórska-Marek, 2011; Zagórska-Marek and Szpak, 2008, 2016; Pennybacker *et al.*, 2015). In discontinuous transitions, the change from one phyllotactic system (m, n) to the next is generally caused by the addition or removal of one or two parastichies in one or both families, for example, from (m, n) to $(m, n - 1)$, $(m - 1, n)$, or $(m - 1, n - 1)$. A discontinuous transition may reverse the chirality of the genetic spiral (Meicenheimer, 1998).

The most common discontinuous transition in plants is believed to be that from the decussate pattern of cotyledons in dicotyledons to the typical spiral pattern of leaves (Meicenheimer, 1998). Elongated cylindrical structures such as the *Magnolia* flower (Zagorska-Marek and Szpak, 2008) or the inflorescence of the Araceae (Jean and Barabé, 2001), which contain a large number of appendages, also show various discontinuous transitions involving different phyllotactic series. For example, the inflorescence of *Anthurium* shows the following transitional sequence: $(17, 22) \rightarrow (14, 22) \rightarrow (13, 20) \rightarrow$ $(12, 19) \rightarrow (12, 18) \rightarrow (12, 17) \rightarrow (11, 16) \rightarrow (10, 16) \rightarrow$ $(10, 15) \rightarrow (10, 14) \rightarrow (9, 14) \rightarrow (8, 14) \rightarrow (8, 13) \rightarrow (6, 12) \rightarrow (7, 6)$ (Jean and Barabé, 2001). However, very little is known about the rare discontinuous transitions from a spiral pattern in seedlings to a decussate pattern, as in *Thuja* (Yin *et al.*, 2011).

Recently, Yin and Meicenheimer (2017) investigated discontinuous phyllotactic transitions in *Diphasiastrum digitatum* (Lycopodiaceae). They measured and compared quantitative parameters including the divergence angle, plastochrone ratio, leaf insertion angle, and van Iterson's (1907) parameter b as well as Douady and Couder's (1996a) parameter Γ across patterns and SAM types. They observed various non-Fibonacci patterns such as $(4, 7)$ and $(5, 9)$ as well as whorled patterns such as $2(1, 1)$, $3(1, 1)$, and $4(1, 1)$, thereby providing counterexamples to Jean's (1994) theoretical model. Discontinuous phyllotactic transitions such as $(5, 8)$ to $(4, 7)$ and $(5, 8)$ to $3(1, 1)$ or $4(1, 1)$ were also seen between SAMs of type V (proximal strobili) and VI (distal strobili). Yin and Meicenheimer (2017) concluded that the divergence angle was the only variable that changed consistently in observed cases of discontinuous transitions; therefore, it could be used to characterise phyllotactic patterns or series. Some (but not all) transitions also showed inconsistent changes in the plastochrone ratio, leaf insertion angle, circumferential ratio, and half conic angle. A discontinuous transition can even involve a change from a regular phyllotactic pattern to a disorganised pattern, especially when the plastochrone ratio (R) and leaf insertion angle (i) are very low ($R < 1.02$, $i < 45°$) and/or when a large number of primordia are initiated in a relatively rapid sequence, as in *Acacia* (Rutishauser, 2016).

Discontinuous transitions have also been studied theoretically (Meicenheimer and Zagórska-Marek, 1989; Jean, 1994; Zagórska-Marek and Wiss, 2003; Yamada *et al.*, 2004; Zagórska-Marek and Szpak, 2008, 2016; Szpak and Zagorska-Marek, 2011; Pennybacker *et al.*, 2015; Douady and Golé, 2016; Golé *et al.*, 2016). Zagorska-Marek and Szpak (2008) developed a geometrical model that enables testing variations in the primordium size in relation to the size of the lateral surface of the SAM. Their simulations showed that for both continuous and discontinuous changes in phyllotaxis, various naturally occurring patterns can be generated based on relatively simple assumptions including changes in the primordium size and consideration of the 'next available' space on a cylinder of constant width. This model considered the fact that more than one initiation site is sometimes possible for an incipient primordium when a shift in the pattern occurs as a result of changing the primordium size and availability of space between primordial units. They concluded that the ultimate selection of a site in conjunction with variations in the extent of change in the size of primordia as they are initiated sequentially impacts the resulting pattern and type of transition. They highlighted this phenomenon in two cases: in *Verbena*, where the pattern transitions from decussate in vegetative shoots to helical in the inflorescence, and in *Magnolia* flowers, where transitions are observed between floral organs of different sizes.

Golé *et al.* (2016) used a disk-stacking iterative model on a cylinder to reproduce phyllotactic irregularities and transitions. By using this model, they explained the presence of irregular transitions in certain patterns. Douady and Golé (2016) used this model for analysing botanical samples. They translated contacts between elements constituting a pattern into graphs and obtained two types of graphs: an ontogenetic graph containing information related to the mode of appearance of primordia and a crystallographic graph which provides information on the overall 'crystallographic' structure. This theoretical method could be used to analyse irregularities in different botanical phyllotactic patterns such as pine cones or artichoke inflorescences.

Shipman and Newell (2005) and Pennybacker *et al.* (2015) generated various types of discontinuous transitions by using a dynamic biomechanical model to elucidate the mechanisms of primordium initiation on the SAM. This approach is discussed in detail in the chapter on "Biomechanical Models."

Variation of Divergence Angle

The measured divergence angle can vary greatly for a given phyllotactic system. In fact, fluctuating divergence angles appear to be a common phenomenon in plants, especially monocotyledons (Hirmer, 1922). For example, the *Euterpe* palm with spiral phyllotaxis shows considerable variation in divergence angles at the SAM level (Barabé *et al.*, 2010). In apical buds of *Chrysalidocarpus lutescens*, the divergence angle varies from 76° to 150° (Fisher, 1973). Kaplan *et al.* (1982) reported that *Chamaeedorea seifrizii* has a spirodistichous phyllotaxis with a mean divergence angle of 148.5° ± 3.9° (range: 134°–163°) and that *Rhaphis excelsa* has a spirodistichous phyllotaxis with mean divergence angle of 143.4° ± 4.2° (range: 135°–153°). *Euterpe oleracea* is characterised by a spiral phyllotactic pattern, and its mean divergence angle per apex varied between 126.9° ± 9.3° and 135.8° ± 8.0°. All divergence angles fluctuated between 115.89° and 157.33° (Barabé *et al.*, 2010).

What is the significance of this variability? For example, is there a correlation between the divergence angle and the plastochrone ratio? In Jean's geometrical model (1994), an increase in the divergence angle is correlated with an increase in plastochrone ratio. According to the FTOP (Adler, 1974; Jean, 1988, 1994), for a given conspicuous parastichy pair value, the divergence angle fluctuates within a predictable interval. The FTOP predicts that for a spiral phyllotactic system characterised by a (2, 3) conspicuous parastichy pair, the divergence angle will be within the interval [1/2, 1/3] 360°, which corresponds to [d = 120°–180°]. Based on the above examples, the ranges of divergence angles agree with the FTOP.

In many plants, changes in divergence angle could be explained by the geometry and growth dynamics of the apex. For example, in developing *Drymis* flowers, the variation of the divergence angle can be predicted and be related to the elliptical shape of the meristem (Doust, 2001). Theoretically, the divergence angle between leaves produced on an ellipsoid apex can oscillate cyclically (Atela *et al.*, 2008). Biologically, it may correspond to a periodic displacement of the histogenetic centre of the apex in conjunction with the initiation of successive primordia. The fluctuating divergence angle could be explained in terms of an auxin sink coming from the last formed leaf (Reinhardt *et al.*, 2003; Kuhlemeier, 2007). This appears plausible because Smith *et al.* (2006b) simulated the fluctuating divergence angle of *Arabidopsis* seedlings by using an auxin-based model.

Hotton (1999) used hyperbolic tessellations to provide rigorous predictions for a dynamical phyllotaxis model. For a given set of parameters, this model can be used to calculate a periodic sequence of divergence angles, for example, {130°, 89°, 89°, 130°, 89°, 89°, 130°, 315°, 130°, …}, with a repeating period of 8. Hotton noted that this theoretical sequence of eight divergence angles agrees fairly well with that observed by Tucker (1961) in

Magnolia flowers: {134°, 94°, 83°, 138°, 92°, 86°, 136°, 310°, 134°, ...}, where divergence angles fluctuate greatly around 137°. This type of model could certainly be used to determine whether this variation occurs randomly or whether it corresponds to a quasiperiodic sequence. Beyer and Rihter-Gebert (2016) mathematically analysed the characteristics of phyllotactic patterns in higher dimensions. They generalised Douady and Couder's two-dimensional model (1996a) to three dimensions. In their simulations, they obtained different behaviours of divergence angles: convergence to a constant value, periodic oscillation, and chaotic fluctuation. However, these sequences have not yet been compared to empirical data. The dynamic phyllotaxis models developed by Atela *et al.* (2002, 2008), Hotton (2008), and Beyer and Rihter-Gebert (2016) for determining rhythmic or pseudorhythmic cycles may be promising for better understanding regular fluctuations in the divergence angle.

Empirical Measures of Phyllotactic Parameters

Parameters can be measured to analyse phyllotactic patterns in various ways. However, all of these approaches aim to maximise the measurement accuracy, especially on microscopic surfaces such as SAMs (Battjes *et al.*, 1993). One of the main problems is the determination of the SAM centre in planar representations. Hotton (2003) presented some methods for finding the centre of phyllotactic patterns. He considered that the main problem faced in finding this point depended on how it was defined. He proposed a practical method for finding the centre, which is defined as the point inside the pattern for which the standard deviation of the divergence angles calculated from this point is the minimum.

Yin and Meicenheimer (2017) used a 'polygon' drawing tool (Paint, Microsoft Windows 7) to connect the flanks and the adaxial midpoint of each primordium at the edge of the SAM. The resulting polygon was smoothed using the program's 'fit spline' function, and the SAM centre was marked as the centroid of the smoothed polygon and assigned Cartesian coordinates (0, 0). The locations of all other points used to measure phyllotactic parameters were noted in relation to this central reference point.

Matkowski *et al.* (1998) developed algorithms to determine with maximal accuracy the middle point of the SAM to measure divergence angles. They discussed two methods to obtain the so-called gravity centre and geometrical centre. In both cases, the positions of primordia (denoted as points) were used as initial inputs. These points are represented in one plane, carry equal weight, and are used to pinpoint the gravity centre. The geometrical centre represents the centre of a series of triangles connecting sequential points with the gravity centre. They determined that using the gravity centre is more suitable and accurate for apices with fewer primordia and an apical dome area that is smaller than the area occupied by lateral primordia. Conversely, using the geometrical centre algorithm is more suitable and accurate for apices with greater diameter, relatively larger meristematic dome, and small primordia (i.e. a higher phyllotactic pattern).

Okabe (2012) proposed a practical method for evaluating the angular coordinates of phyllotactic units based on the concept of a floating centre. This method does not assume any specific functional form of the distance of the element to the centre of the apex nor the existence of a fixed centre for the pattern. Okabe (2015a) used this method to precisely calculate the divergence angle of *H. annuus* capitula and the number of florets in the inflorescence.

He showed that the floating centre can be the main source of apparent variations in divergence angles.

In palaeobotany, plant fossils are often incomplete, and determining phyllotactic patterns in these structures remains very problematic. To characterise the phyllotaxis of isolated fragments, such as those of *Lepidodendron*, Desmet and Schaaf (1995) proposed a spectral analysis method that integrated all topological parameters of the pattern appearing on the fossil.

Disorganised Phyllotactic Patterns

Most theoretical models are based on regular patterns, generally represented by lattices, appearing at the SAM level. However, disorganised, or aberrant (Jean, 1994), phyllotactic patterns occur naturally in shoots (Barabé, 2006; Rutishauser, 2016). Such patterns also appear in many mutants studied in molecular biology (e.g. Itoh *et al.*, 2000). Genetic mutations can induce unpredictable variations in phyllotactic parameters; some modify the divergence angle (*pin-formed 1-1, revoluta*) whereas others modify the phyllotactic system, for example, from whorled phyllotaxis to spiral phyllotaxis (*floricaula, squamosa*) (see Appendix 1 for a more detailed list of mutants in the chapter on "Role of Genes in the Framework of Biochemical and Molecular Models"). The mutations involved in phyllotaxis mainly affect the size of the foliar primordia or the apex. For example, wild-type maize is characterised by a distichous phyllotaxis. However, in the mutant *abphyl*, two leaves are produced in opposite-decussate positions (whorl of two leaves) (Jackson and Hake, 1999). The SAM of the mutant has a larger size than that of the wild-type but has unaltered shape. The *fey* mutation causes a change in the divergence angle and the SAM structure (Callos *et al.*, 1994) of *Arabidopsis*. In this case, there is no correlation between the divergence angle and the plastochrone ratio, contrary to the predictions of theoretical models (Jean and Barabé, 1998b). In the rice mutant that shows the generation of random divergence angles, modified plastochrone ratio, and misshapen leaves and SAM (Itoh *et al.*, 2000), no statistical correlation is seen between parameters (Jeune and Barabé, 2006b).

Although developments in molecular biology can help in better understanding phyllotaxis, it remains difficult to associate a single mutation with a single phyllotactic parameter because a given gene can be simultaneously involved in controlling the position and form of leaves and the SAM. Unlike with regular patterns, there is no correlation between phyllotactic parameters in disorganised phyllotactic patterns.

Conclusion

The value of phyllotactic parameters in the development of models and in experimental studies cannot be understated. This is highlighted by subsequent chapters that discuss various approaches and studies related to pattern generation in plants. The present chapter primarily focused on the foundational geometrical phyllotactic parameters used to describe and analyse phyllotactic patterns. However, with the development of new phyllotaxis models such as biomechanical and biochemical models, additional parameters linked specifically to the functioning of the SAM need to be considered to more completely explain the emergence of patterns in plants.

2 Dynamical Models

Phyllotactic studies have led to the development of various models to simulate/emulate naturally occurring patterns. This chapter explores the evolution of these models and focuses on energetic models involving minimum energy principles or chemical inhibitory fields.

Theoretical analyses of phyllotaxis mainly deal with the modelling of regular phyllotactic patterns by using deterministic approaches, such as the behaviour of dynamical systems involving the interaction of chemical inhibitory fields (e.g. Bernasconi and Boissonade, 1997; Koch *et al.*, 1998; Kunz, 2001; Meinhardt, 2003, 2004; Smith *et al.*, 2006a). Other deterministic analyses involve, for example, energy principles (e.g. Levitov, 1991a,b; Douady and Couder, 1992, 1996a–c; Cummings and Strickland, 1998; Guerreiro and Rothen, 1998; Kappraff *et al.*, 1998; Lee and Levitov, 1998; Marzec, 1998, 1999a,b; d'Ovidio *et al.*, 1999; d'Ovidio and Mosekilde, 2000; Atela *et al.*, 2002; Shipman and Newell, 2004), density of structures packed into a theoretical space (e.g. Malygin, 1998, 2000; Bryntsev, 2000; Hauk and Mika, 2003; Pennybacker and Newell, 2013; Mughal and Weaire, 2017), contact pressure (Adler, 1998; Hellwig *et al.*, 2006), topological defects in a continuous medium (Miri and Rivier, 2002; Rivier *et al.*, 2016), dendritic growth in physics (Fleury, 1999), or numerical and geometric properties of phyllotactic organisation (Jean, 1994, 1998b; Wilson, 1995; Bursill and Rouse, 1998; Kappraff *et al.*, 1998; Van der Linden, 1998; Petrov, 2001; Takaki *et al.*, 2003; Yeatts, 2004; Hellwig and Neukirchner, 2010; Bergeron and Reutenauer, 2019). A group-theory-based approach[1] which is currently used in physics and mathematics (Yamada *et al.*, 2004) can also reproduce different phyllotactic patterns. From a topological viewpoint, Rivier *et al.* (2016) analysed the organisation of phyllotactic patterns by comparing it to a geometrical form in which a large disk is densely covered by Voronoi cells (leaves or florets) of the spiral lattice. They showed that this representation into parastichies and grain boundaries is not sensitive to cell disappearance and can accommodate shear stress naturally. Their geometrical model agrees well with the phyllotaxis of *Agave parryi*.

Another deterministic approach used to analyse the phyllotactic patterns involves light capture efficiency. This problem has been studied empirically (Valladares *et al.*, 2002; Zotz *et al.*, 2002; Valladares and Brites, 2004) and theoretically (Sekimura, 1995; Niklas, 1998; King *et al.*, 2004) as well as from an evolutionary perspective (Altesor and Ezcurra, 2003).

This chapter on "Dynamical Models" briefly discusses modern approaches used to generate the phyllotactic patterns.

Chemical Models

The underlying mechanisms involved in regulating phyllotaxis have been studied extensively in the past through surgical experiments (Loiseau, 1969; Steeves and Sussex, 1989; Reinhardt *et al.*, 2005)

---

[1] Group theory is a mathematical discipline involving studying algebraic structures called groups. A group is a set of elements in which any two elements can be combined to obtain a third element. Group theory is used to study symmetrical structures such as crystals and molecules and the spectroscopic properties of different materials.

or by using growth regulators (Wardlaw, 1949a,b; Snow and Snow, 1962; Charlton, 1974; Meicenheimer, 1981). Schoute (1913) first proposed the involvement of a chemical substance in the formation of phyllotactic patterns and hypothesised that a new primordium produces a diffusible chemical substance that inhibits the initiation of primordia in its immediate neighbourhood. Richards (1948) and Wardlaw (1949a) later elaborated this theory based on empirical data. Turing (1952) then introduced the use of dynamical systems to model different biological phenomena including phyllotaxis. However, theoretical models based on the diffusion of one or two substances (inhibitor and/or activator) were developed only in the 1970s (Hellendoorn and Lindenmayer 1974; Thornley, 1975; Mitchison, 1977; Veen and Lindenmayer, 1977; Richter and Schranner, 1978; Young, 1978; Meinhardt, 1982, 2003; Schwabe and Clewer, 1984; Chapman and Perry, 1987; Yotsumoto, 1993; Meinhardt *et al.*, 1998; Smith *et al.*, 2006a). Interestingly, models based on the activity of an inhibitor were the first to incorporate computer science techniques to generate phyllotactic patterns. The model of Smith *et al.* (2006a) represents a relatively recent and good example of inhibition by a diffusible substance.

Smith *et al.* (2006a) generated different phyllotactic patterns by using a simple model based on inhibitory fields surrounding existing primordia. They used inhibition functions that depend on the spatial distribution and age of existing primordia. In this model, the critical parameter is the inhibition threshold at which new primordia are formed. The inhibitory field itself is calculated by using the existing primordia around the shoot apical meristem (SAM) (Fig. 2.1). They analysed two simulation models: one with a single inhibition function that generates patterns where primordia are initiated successively and another with two inhibition functions where two or more primordia appear simultaneously.

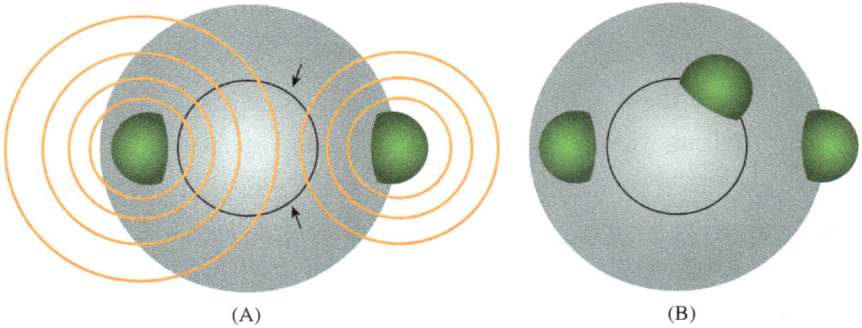

(A)   (B)

Fig. 2.1. Diagram of inhibition. **(A)** The older primordium has a smaller inhibiting effect on the active ring than the newer primordium. Arrows indicate minima of inhibition where a new primordium can appear. **(B)** The upper minimum inhibition was chosen as the location of the third primordium. (From Smith *et al.*, 2006a; reprinted with permission from Canadian Science Publishing.)

In the model with a single inhibition function, the total inhibiting effect $h(S)$ of n previously formed primordia is given as

$$h(S) = \sum_{i=1}^{n} \frac{1}{d(P_i, S)} e^{-bt_i} \tag{1}$$

where $d(P_i, S)$ is the distance between primordium P_i and a sampling point S on the surface of the apex; b, the parameter controlling the rate of exponential decrease in inhibition over time; and t_i, the age of primordium i.

By changing the value of the inhibition threshold, which controls the position of newly formed primordia, Smith *et al.* (2006a) generated different spiral patterns. Relatively low thresholds correspond to a distichous pattern. Increasing the threshold led to a switch to spirodistichous and then to Fibonacci patterns (Fig. 2.2). They also generated a similar series of phyllotactic patterns by keeping the threshold value constant and varying the size of the apex. This result concurs with morphological observations showing a correlation between the size of the apex and the type of

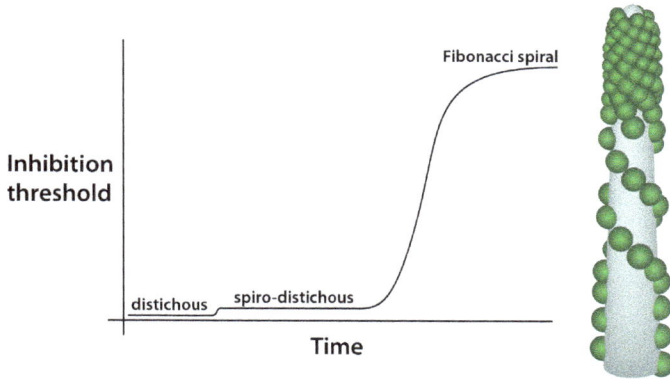

Fig. 2.2. Effects of changes in inhibition threshold. At low inhibition thresholds, a self-starting distichous phyllotaxis is produced. Increases in threshold values cause a switch to spirodistichous and then to spiral Fibonacci patterns. (From Smith *et al.*, 2006a; reprinted with permission from Canadian Science Publishing.)

phyllotactic pattern (Rutishauser, 1998; Jackson and Hake, 1999; Kwiatkowska and Florek-Marwitz, 1999).

A simulation model with two inhibition functions (i.e. two inhibitors) was developed to generate spiral and whorled phyllotactic patterns. Smith *et al.* (2006a) modified the preceding equation to express long-range inhibition that determines the phyllotactic organisation of primordia:

$$h_l(S) = (1.0 - e^{-b_d t_i}) \sum_{i=1}^{n} \frac{1}{d(P_i, S)} e^{-b_l t_i} \qquad (2)$$

where the additional parameter b_d controls the rate at which the inhibiting influence of the primordium is introduced and b_l, the exponential decay of inhibition. According to Eq. (2), newly initiated primordia do not necessarily produce inhibition fields immediately at other locations, thereby facilitating the initiation of other primordia concurrently at other minima on the SAM. This process allows for the emergence of whorled and multijugate patterns. Smith *et al.* (2006a)

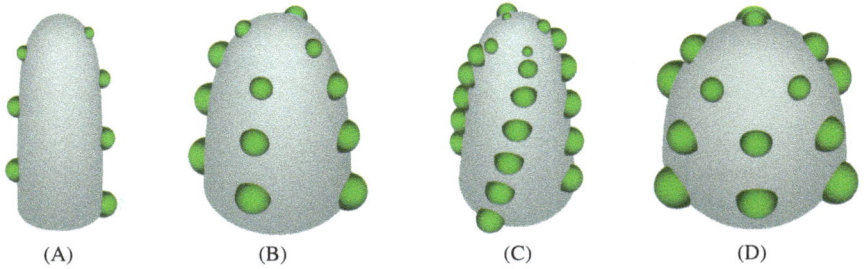

Fig. 2.3. Examples of phyllotactic patterns produced by the two-inhibitor model: **(A)** distichous, **(B)** decussate, **(C)** spirodecussate, and **(D)** whorled. (From Smith *et al.*, 2006a; reprinted with permission from Canadian Science Publishing.)

used another equation to express short-range inhibition to prevent the close formation of multiple primordia on the same active ring:

$$h_s(S) = \sum_{i=1}^{n} \frac{1}{d(P_i, S)} e^{-b_s t_i} \tag{3}$$

where the new parameter b_s controls the rate of exponential decay of the second inhibitor. Consequently, the positioning of primordia remains under the control of long-range inhibition because the effect of short-range inhibition decreases comparatively faster ($b_s > b_l$).

By using this simple two-inhibitor model function (Eqs. (2) and (3)), Smith *et al.* (2006a) simulated distichous, spirodistichous, spiral and whorled patterns (Fig. 2.3).

Energy-based Dynamical Models

Energy-based dynamical models consist of systems in which dynamical self-organising processes in the form of repulsive energy are used to reproduce the phyllotactic patterns.

Levitov's study (1991a,b) on the phyllotaxis of flux lattice in superconductors and Douady and Couder's study (1992, 1996a–c, 1998) on the self-organising processes involved in phyllotaxis are based on the idea that phyllotactic patterns minimise repulsive energy among neighbouring primordia. These studies marked the beginning of a new generation of phyllotactic models involving dynamical processes instead of purely geometrical ones. Douady and Couder (1998) considered that phyllotaxis is a self-organising phenomenon and that phyllotactic patterns can be generated by a set of simple dynamical rules independent of the biological or physical processes responsible for producing primordia. These new dynamic phyllotactic models provide mathematical explanations for the mode of appearance and stability of phyllotactic patterns linked to Fibonacci numbers (Atela *et al.*, 2002) and show that the emergence of phyllotactic patterns represents a dynamical self-organising process that occurs owing to local repulsive interactions between primordia that form during the growth of the apical meristem. Physical phenomena reproducing phyllotactic patterns comparable to those reported in experiments with ferromagnetic fluid (Douady and Couder, 1992) were also analysed on the surfaces of self-assembled spherical microstructures (diameter: ≈10 µm) (Li *et al.*, 2005), an experimental magnetic cactus (Nisoli *et al.*, 2009), and bubbles emerging periodically from a liquid surface (Yoshikawa *et al.*, 2010).

In their theoretical models, Douady and Couder (1996a–c) simulated the dynamic nature of the SAM to emulate the growth of real plants. Later, several mathematicians and physicists produced models and analyses dealing more precisely with the theoretical structure of the dynamical system than their plausibility in a botanical context.

Hotton *et al.* (2006) formally showed that many dynamical models (Douady and Couder, 1996a–c; Kunz, 1997; Koch *et al.*, 1998; d'Ovidio and Mosekilde, 2000; Atela *et al.*, 2002) can be considered discrete dynamical systems using points

$(R_0, \theta_0), \ldots, (R_N, \theta_N)$ on the disk. Each point represents the center of a primordium in polar coordinates. The transformation from one configuration to the next corresponds to the following formal relation:

$$(R_0, \theta_0, \ldots, R_N, \theta_N) \rightarrow (f(R_0, \theta_0, \ldots, R_N, \theta_N), R_0, \theta_0, \ldots, R_{N-1}, \theta_{N-1}),$$

where the function f determines the position of a new primordium that minimises its interaction with existing ones. This corresponds to a location and time at which interaction W decreases below a given threshold. In most models, the interaction is given by the following function:

$$W(R, \theta, R_0, \theta_0, \ldots, R_N, \theta_N) = \sum_k u(R, \theta, R_k, \theta_k),$$

where (R, θ) is the 'test' location of a new primordium and u, a function linked to the distance between (R_k, θ_k) and (R, θ). In their model, Douady and Couder (1996a) used $u = c/\text{dist}^a$, where a and c are constants and dist is the distance between (R_k, θ_k) and (R, θ). The strength of the interaction affects the type of phyllotactic pattern that will be produced.

Atela *et al.* (2002) extended Hotton's study (1999) and presented a rigorous mathematical analysis of a discrete dynamical model based on Douady and Couder's study (1996a–c). They explained the frequent occurrence of the Fibonacci sequence in the number of parastichies by the stability of the fixed points in this system as well as the structure of their bifurcation diagram (also called phyllotactic tree). Hotton (1999) noted that Kunz (1995, 1997) independently developed a similar dynamical system. Kunz (1995) presented mathematical results for generalising Levitov's phyllotactic model (1991a,b) and Douady and Couder's experiment (1992), which were essentially equivalent (Guerreiro and Rothen, 1998). He showed that the hierarchical selection of noble numbers in phyllotactic models was closely linked to the two-dimensional geometry of the system. Recently, Atela (2011) developed a dynamic geometrical model showing how a small change in the position of the first-formed primordium can lead to different phyllotactic systems: Fibonacci, Lucas, bijugate, and multijugate. However, although these models remain very theoretical, some of them provide new insights to explain the presence of phyllotactic patterns in many natural systems and highlight new results that could help in analysing empirical data.

Recently, Yonekura *et al.* (2019) extended the Douady and Couder model to generate what they call orixate phyllotaxis as observed in *Orixa japonica*. The orixate phyllotaxis corresponds to a tetrastichous alternate pattern characterised by the periodic repetition of a sequence of different divergence angles.

Given that Douady and Couder's model represented a paradigm shift for certain researchers (Shipman and Newell, 2005; Mirabet *et al.*, 2012) and inspired many theoretical analyses (e.g. Liaw, 1998; Hotton, 1999; d'Ovidio and Mosekild, 2000; Atela *et al.*, 2002; Liu and Liaw, 2007; Shipman *et al.*, 2011; Beyer and Richter-Gebert, 2016) and experimental studies (Barabé *et al.*, 2007, 2009), we discuss the main characteristics of this model.

Douady and Couder's first iterative dynamical model (1996a) used Hofmeister's (1868) principle of the largest available space. Douady and Couder (1996b) noted that this system can only give rise to distichous and spiral patterns. To incorporate whorled and multijugate patterns, Douady and Couder developed a second model based on Snow and Snow's developmental rules (1952), which state that a new primordium will form where and when there is enough space for its formation. In their simulation, Douady and Couder (1996b) used a planar representation of a convex apex and formulated a set of algorithmic rules to construct their dynamical iterative model in this representation.

(A)

(B)

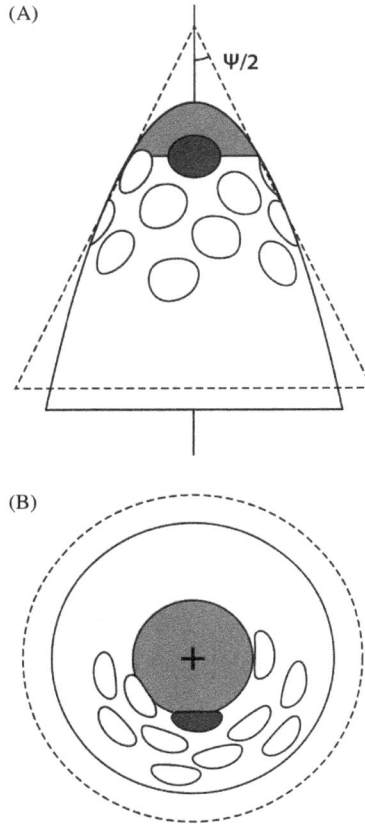

Fig. 2.4. **(A)** Scheme showing the apex and the cone tangent to it in the region of formation of primordia (dashed lines). The cone is characterised by its angle $\psi/2$. **(B)** The projection in the transverse section plane. The folioid shape of the elements results from their projection. The drawings correspond to the case in which the sizes of primordia do not change as they move away from the apex. (From Douady and Couder, 1996b; reprinted with permission from Elsevier.)

Primordia are represented by circles with diameter d_0 on a convex apex (paraboloid) with the curvature represented by a parameter N. If $N = 1$, the apex is flat; if $N > 1$, it has a curvature. When projected in a plane perpendicular to the SAM, circular primordia will appear azimuthally elongated (Fig. 2.4). In a planar representation, the apex is circular with radius R; outside this region, no further reorganisation leads to changes in the angular positions of primordia. Primordia are initiated at the periphery of the apex and move away or are left behind as a result of growth with constant radial velocity V. Each primordium generates repulsive energy $E(d)$, where d is the distance from the new primordia. A new primordium is initiated on the apex at a specific location and time such that the sum of the repulsive energy generated by all previous primordia becomes smaller than a chosen threshold E.

On the projection plane, the distance between two points $P_0 = (r_0, \theta_0)$ and $P_1 = (r_1, \theta_1)$ is described in polar coordinates as follows:

$$d(P_0, P_1) = \sqrt{\left[\frac{r_0^2 + r_1^2}{N} + 2Nr_0r_1 (1 - \cos(\theta_0 - \theta_1)) \right]}. \qquad (4)$$

To calculate the repulsive energy that a primordium exerts on a point (future primordium) located at distance d from it, Douady and Couder used the following interaction law:

$$E(d) = \frac{-1 + \left(\tanh \alpha \, \dfrac{d}{d_0} \right)^{-1}}{-1 + (\tanh \alpha)^{-1}} \qquad (5)$$

where $\tanh \alpha$ is the hyperbolic tangent and α, the interaction force (stiffness). The authors also noted that the following simpler equation gave similar results:

$$E(d) = \frac{1}{(d/d_0)^p} \qquad (6)$$

where p is the stiffness of the interaction. In both equations, the primordium size d_0 is the main controlling parameter of the system. The threshold potential $E_s = E(d_0)$ is always equal to 1 for $d = d_0$.

For lower α values, a primordium has a larger influence on another. The energy $E(\theta)$ exerted on a primordium (point) on the generative circle (Fig. 2.4) is defined as the sum of the repulsive energies exerted on the point by all previously formed primordia. Of course, the energy of primordia decreases with their increasing distance from the SAM centre.

When leaves are represented by circular disks on a plane, the number of contact parastichies and conspicuous parastichies is the same and the angle of intersection between opposite parastichies is closest to 90°. However, with folioid-shaped elements, the number of contact parastichies and conspicuous parastichies differs. Botanists typically use contact parastichies in their empirical studies. However, some theorists, mostly mathematicians (Jean, 1994), base their phyllotactic systems on the concept of conspicuous parastichies because such systems do not consider the form of primordia. Nonetheless, in Douady and Couder's model (1996b, p. 277), '... the dynamically important spirals which govern the selection of the pattern are the contact parastichies. This is related to the fact that the contact parastichies of the folioids in the plane correspond to parastichies that are both contact and nearest neighbour spirals on the initial conical surface'.

In Douady and Couder's model (1996b), a geometrical parameter $\Gamma = d_0/R_0$, where d_0 is the diameter of the primordium and R_0 is the radius of the apex, controls the dynamics of the system. This parameter is similar to $b = d_0/2\pi R_0$ used by van Iterson in his geometrical analysis of the phyllotactic organisation. Douady and Couder (1996a,b) showed that Γ can be linked to the shape of folioids (N, planar representation of primordium) and conicity ($1/N$), where $N = \sin(\Psi/2)$, the angle of the cone, as $\Gamma = d_0/N_1/2R_0$. By changing Γ, they obtained a phase diagram representing different spiral or whorled patterns. Interestingly, neither the periodicity (plastochrone) nor divergence angle is prescribed in this model; they emerge from the self-organisation of an iterative system.

In Douady and Couder's model, a continuous selection of divergence angles (180°–137°) exists when starting with a low number of primordia. However, a large number of divergence angles are seen when starting with a high number of primordia (Couder, 1998). Douady and Couder (1998) noted that when Γ becomes very small, the divergence angle can be very different from that predicted by the theoretical model. For example, instead of the divergence angle remaining constant at a given value, the system shows anomalous values such as 137.5°, 275°, 222.5°, and 275°. This variation corresponds to a permutation in the order of appearance of two successive elements. However, Douady and Couder (1998) noted that although there are permutations in the order of appearance of the elements and, consequently, in the divergence angle, parastichies are still easy to observe. They opined that even if the regularity of the iteration is locally perturbed, it weakly affects the stability of spiral patterns.

Newell and Shipman (2008) noted that Levitov's (1991a,b) and Douady and Couder's (1992, 1996a–c) models do not consider the form of primordia and do not explain the phyllotactic tiling planform (e.g. hexagonal, parallelogram). In models based on repulsive energy between primordia, the appearance of a phyllotactic pattern is independent of the physical or biological mechanisms responsible for producing new primordia. The central idea is that phyllotactic patterns will appear as long as each new primordium is initiated following a given rule in terms of energy between repelling primordia in a space where it can optimise its growth irrespective of the form of its neighbours.

Empirical Observations and Dynamical Models

One of the main challenges in phyllotaxis is the quantitative comparison between theoretically simulated and observed patterns (Battjes and Prusinkiewicz, 2008) to determine the accuracy of a model. To address this issue, Hotton *et al.* (2006) proposed a new concept called ontogenetic graph to compare observed and simulated patterns. The concept of an ontogenetic graph is based on the mode of development of a particular pattern and provides an original approach to compare naturally occurring patterns with those generated by theoretical models. They showed how a simple phyllotactic model can generate common naturally occurring phyllotactic patterns. The geometrical dynamical model of Hotton *et al.* (2006) is derived from a previous study by Atela *et al.* (2002). The iterative model is based on two rules (p. 315): '1. Primordia are formed in succession, one or more at a time. 2. Primordia are positioned in the least crowded spot at the edge of the meristem'.

Further, Hotton *et al.* (2006) developed a dynamical system involving an iterative process based on the concept of the ontogenetic graph that can reproduce observed phyllotactic patterns. The ontogenetic graph of a pattern corresponds to the vertices joining each primordium to its closest older neighbouring primordia. The ontogenetic graph considers the topology and ontogeny of a pattern. They applied their model to *Helianthus annuus* and *Cynara scolymus* capitula. They used stereoscopic reconstructions of different stages of development of the capitula to compare observed and simulated patterns (Fig. 2.5). They found a close fit between the observed patterns and those generated by a simple dynamical model.

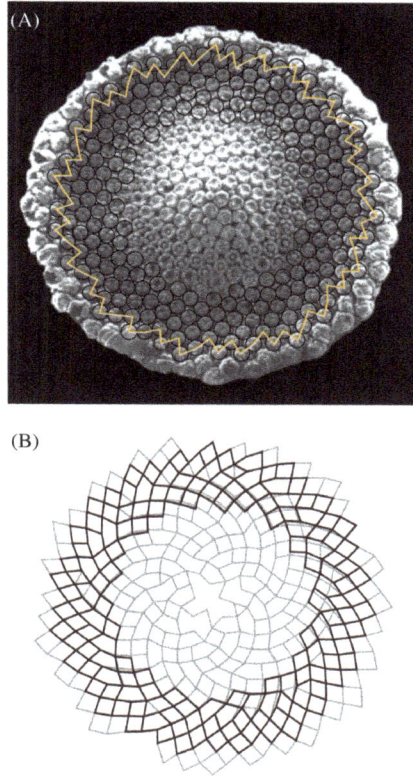

Fig. 2.5. Dynamical simulation of artichoke capitulum using the constant radius ($r = 0.095$ mm, 150 iterations are shown). (**A**) SEM image of capitulum with simulated primordia superimposed in black. The initial form is represented by a black zigzagging loop. (**B**) The ontogenetic graph of the simulation (black) superimposed on that of the data (grey). (From Hotton *et al.*, 2006; reprinted with permission from Springer Nature.)

Wave Model of Phyllotaxis Involving Auxin

Abraham-Shrauner and Pickard (2011) proposed a phyllotactic model involving the propagation of oppositely directed waves on the periphery of a theoretical generative circle. The auxin concentration reaches its maximum at the point where the waves meet, resulting in the initiation of a foliar primordium. Although their model is based on auxin transport, the mode of control of auxin transport differs from that reported previously. In their model, waves move out azimuthally from a young primordium in opposite directions (clockwise and counterclockwise). When these two waves collide, a new primordium is formed. In this model, the position of a new primordium depends on the wave speed and the timing between the wave fronts. Further, the plant stem is represented by a cylinder (Fig. 2.6), and the coordinates are the radius r, azimuthal angle ϕ, and axial coordinate z.

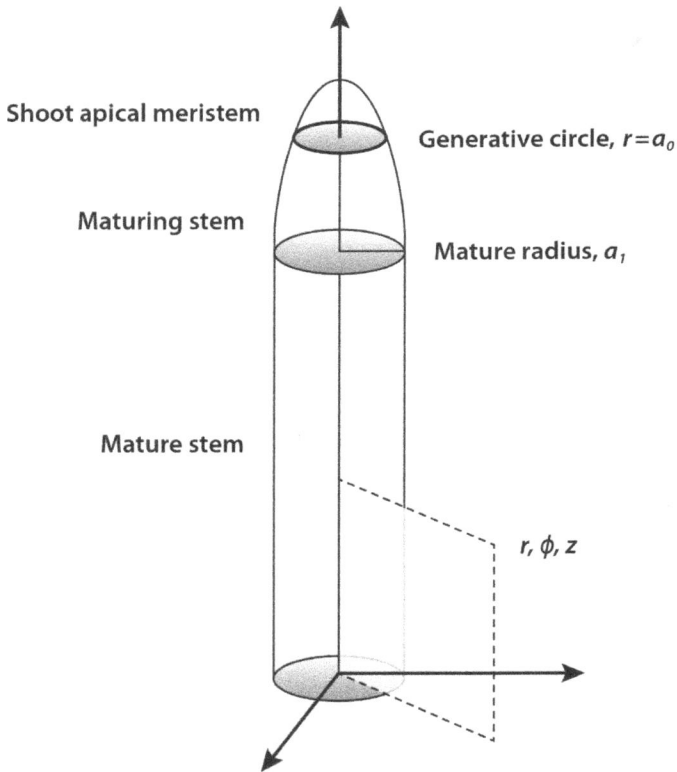

Shoot apical meristem

Generative circle, $r=a_0$

Maturing stem

Mature radius, a_1

Mature stem

r, ϕ, z

Fig. 2.6. An abstraction of the plant stem. Key geometric features of the mathematical wave model are shown; in particular, the SAM and the generative circle within it are indicated on top of the stem. Circular cylindrical coordinates are defined. (From Abraham-Shrauner and Pickard, 2011; reprinted with permission from Taylor & Francis.)

The mature stem has a radius a_1, and the leaf primordia are initiated on a generative circle of radius a_0 with divergence angle ϕ_d. Corresponding to this geometrical representation, the divergence angle ϕ_d and its complement ϕ_c (in radians) for spiral phyllotaxis are

$$\phi_d = \frac{M_s\, 2\pi}{N} \tag{7A}$$

$$\phi_c = \frac{M_f\, 2\pi}{N} \tag{7B}$$

where M_s is the smaller number of turns between specific leaves in a sequence; M_f, the larger number of turns; and N, the number of leaves in the sequence. The number of turns may vary depending on whether the rotation is clockwise or counterclockwise. From Eqs. (7A) and (7B), the authors obtained

$$M_S + M_f = N. \tag{8}$$

This relationship corresponds to a Fibonacci series in which N is the sum of the two previous terms. For whorled phyllotaxis patterns, the divergence angle between whorls is given by

$$\phi_d = \frac{2\pi}{N} \tag{9}$$

where L is the number of leaves and $N = 2L$.

Wave equations

The two waves A and B are respectively slow and fast and correspond to different auxin concentrations. This dynamical system is expressed by the following equations:

$$\nabla^2 A - \frac{1}{V_s^2}\frac{\partial^2 A}{\partial t^2} = \frac{1}{a_0^2}\frac{\partial^2 A}{\partial \phi^2} - \frac{1}{V_s^2}\frac{\partial^2 A}{\partial t^2} = 0 \tag{10A}$$

$$\nabla^2 B - \frac{1}{V_f^2}\frac{\partial^2 B}{\partial t^2} = \frac{1}{a_0^2}\frac{\partial^2 B}{\partial \phi^2} - \frac{1}{V_f^2}\frac{\partial^2 B}{\partial t^2} = 0 \qquad (10B)$$

where V_s and V_f are respectively the speed of the slow and fast wave, and a_0 is the radius of the generative circle or the level at which primordia are initiated on the SAM.

The solutions for these equations must present a periodicity for the azimuthal angle ϕ and time t. The authors considered that the divergence angle ϕ_d and its complement ϕ_c are related to the wave speeds (V_s and V_f) and angular frequencies (ω_s and ω_f) by the following equations:

$$\phi_d = \frac{V_s \tau}{a_0} = \omega_s \tau \qquad (11A)$$

$$\phi_c = \frac{V_f \tau}{a_0} = \omega_f \tau \qquad (11B)$$

where τ is the time required for the wave fronts to move from the leaf primordium to the point where they meet on the generative circle.

From particular solutions to Eqs. (10A) and (10B), Abraham-Shrauner and Pickard (2011) generated spiral and whorled phyllotactic patterns. In spiral phyllotaxis (except for a distichous pattern), two waves propagate in opposite directions at different angular speeds. In whorled phyllotactic patterns, two waves with the same angular speed propagate from each of the first two leaf primordia for an opposite-decussate pattern and from each primordium for multiprimordial whorled patterns.

Although Abraham-Shrauner and Pickard's model (2011) considers the auxin concentration, it is highly hypothetical from a biological viewpoint. This model is more descriptive rather than explanatory for the phyllotactic organisation.

Gradient Flow of Auxin

Pennybacker and Newell (2013) derived a new model from a continuous approximation of the discrete biochemical models of Barbier de Reuille *et al.* (2006) and Jönsson *et al.* (2006). In their model, the instability of uniform auxin concentration due to reverse diffusion leads to the appearance of phyllotactic patterns. Reverse diffusion occurs when PIN1 proteins move from the inside of the cell to the cell wall and guide auxin flow from a higher to a lower concentration. They assumed that surface deformation depends on the auxin field concentration and did not consider non-gradient–related variables in the biochemical equation. The resulting gradient flow can be described by the following partial differential equation (PDE; Newell and Pennybacker, 2013; Pennybacker and Newell, 2013; Pennybacker *et al.*, 2015):

$$\frac{\partial u}{\partial t} = \frac{\delta \varepsilon}{\delta u} = \mu u - (\nabla^2 + 1)^2 u - \frac{\beta}{3}(|\nabla u|^2 + 2u\nabla^2 u) - u^3. \tag{12}$$

Further, the energy is given by

$$\varepsilon[u] = -\int \frac{\mu}{2}u^2 - \frac{1}{2}(u + \nabla^2 u)^2 + \frac{\beta}{3}u|\nabla u|^2 - \frac{1}{4}u^4. \tag{13}$$

In these equations, u corresponds to the fluctuation of auxin concentration around a mean value with time (t). The authors scaled the system so that the most linearly unstable wavelength is 2π. u represents the amount by which reverse diffusion through the activity of PIN1 transport overcomes ordinary diffusion and other 'loss effects'. β is a measure of auxin concentration dependent on PIN1 distribution. Based on this equation, phyllotactic patterns are generated by a pushed wavefront moving on a discoid surface.

Newell and Pennybacker (2013) and Pennybacker and Newell (2013) studied numerical solutions to Eq. (13) in the context of a discoid geometry corresponding to a sunflower head. They showed that the local energy density is minimal and packing efficiency is maximal when circumferential wave numbers (i.e. a number of parastichies; e.g. 89, 144, and 233) belong to the Fibonacci series and that hexagonal patterns are more stable than rhombic patterns.

They also noted a relationship between the rise $\rho\,(r)$ (distance between two successive levels of configuration points) and the corresponding divergence angle $\delta(r)$ at each radius. They obtained a Fibonacci branch; the maxima of the pattern formed by a push front governed by Eq. (1) coinciding almost exactly with that is shown in van Iterson's phase diagram (Fig. 2.7).

Interestingly, the travelling-wave-based continuous approach (Eq. (1)) used by Pennybacker and Newell (2013) gives results

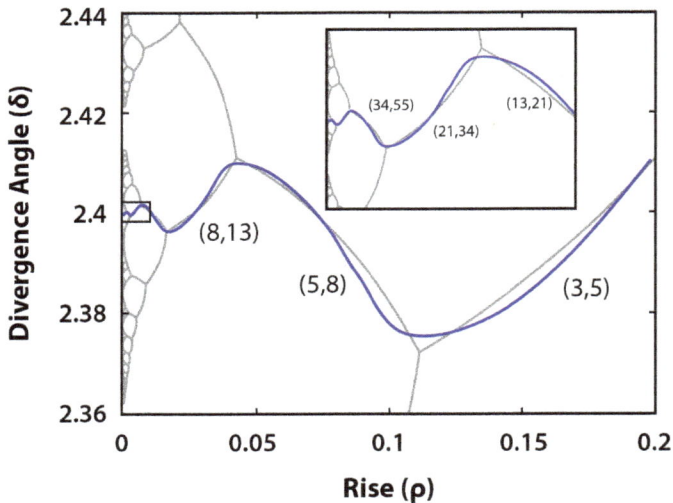

Fig. 2.7. Rise ρ and divergence angle δ (in radians) given by the local lattice structure at each radius on the (1, 2) sunflower. The shaded lines indicate van Iterson's tree, with selected parastichy pairs indicated. The inset shows details of the data for small ρ. A movie may be found at http://math.arizona.edu/~pennybacker/media/divergence/. (From Newell and Pennybacker, 2013; reprinted with permission from Elsevier.)

similar to van Iterson's geometrical-parameter-based discrete approach. Their results are also consistent with Douady and Couder's (1996a–c) and Levitov's (1991a,b) results.

Conclusion

The models discussed in this chapter are based on the principle of repulsion or inhibition between primordia or the propagation of a wave corresponding to the concentration of hormones. However, they do not consider the physical or biological processes and parameters involved at the SAM level. During the 2000s, models involving biochemical processes acting at the cellular level or physical processes acting at the SAM surface were developed separately. These models are based on different molecular and physical parameters and are discussed in greater detail in Chapters 4 ("Role of Genes in the Framework of Biochemical and Molecular Models") and 5 ("Biophysical Aspects of Phyllotaxis").

3 Statistical and Probabilistic Approaches

Theoretical phyllotaxis models are generally based on geometric regularities appearing at the SAM level. Botanists, chemists, physicists, and mathematicians have long focused on patterns that can be generated using self-organisation rules. However, the divergence angle, one of the main parameters used to characterise phyllotactic patterns, is not always fixed and can vary around an average value. In some plants, the amplitude of variation of the divergence angle is so large that it leads to the appearance of irregular phyllotactic patterns. This phenomenon was observed frequently in wild plants (e.g. Loiseau, 1969; Endress, 1989; Jean, 1994; Barabé *et al.*, 2010), in mutants of *Arabidopsis* (Peaucelle *et al.*, 2007, 2011; Ragni *et al.*, 2008; Prasad *et al.*, 2011), and rice in which an initial regular pattern becomes perturbed in *shoot organisation* (*sho*) mutants (Itoh *et al.*, 2000) (Fig. 3.1).

In mutants, the term 'perturbed' (Jean, 1994) can be used to describe irregular sequences of angles. This term is related to particular growth conditions that lead to roughly disordered patterns. Jean (1994) stated that 'When the perturbation is small, the mechanism is self-correcting and confines the divergence angle around a certain value. When it is a bit stronger, [...] the perturbations produce a shift to another standard pattern with an eventual reversion back to the initial one'. The biological processes involved in perturbed patterns may be the same as those involved in deterministic models. However, the extent of their quantitative

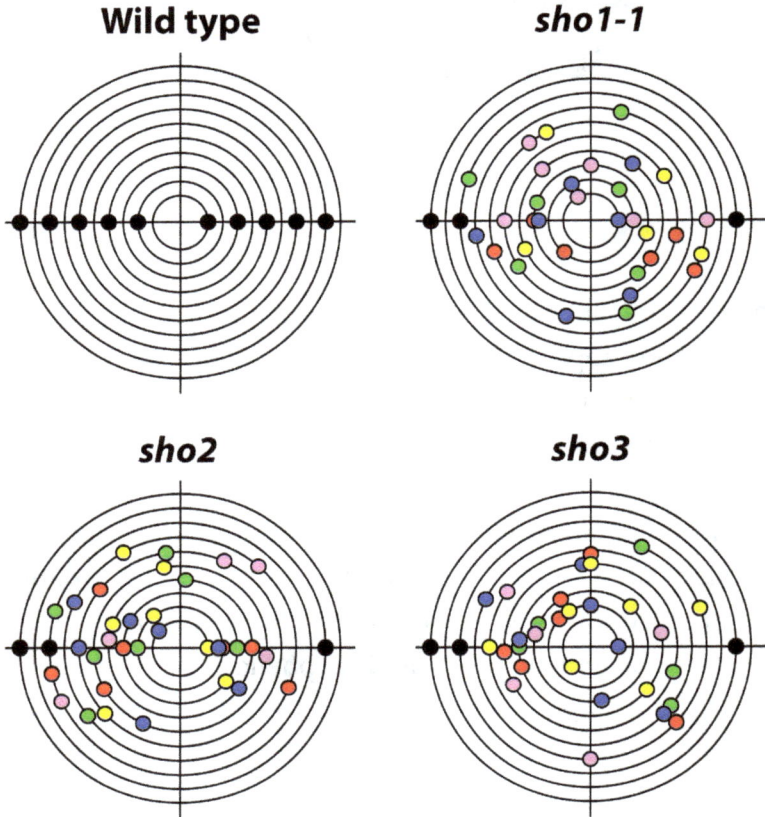

Fig. 3.1.   Schematic representation of phyllotactic pattern in *sho* mutants. Positions of the first through tenth leaf primordia are indicated by coloured circles on concentric circles: the outermost circle corresponds to the first leaf; the innermost circle, to the tenth leaf; and the centre, to the SAM. Different colours indicate different seedlings. In the wild type, leaves are formed in distichous phyllotaxis, as indicated by black circles. The positions of the first through third leaves in *sho* mutants are almost normal and are indicated by black circles. (From Itoh *et al.*, 2000; reprinted with permission from American Society of Plant Physiologists.)

variation remains unpredictable. Many genetic mutations involved in meristem organisation can modify the values of phyllotactic parameters. For example, some mutations such as *pin-formed 1-1* (Reinhardt *et al.*, 2000) and *Revoluta* (Otsuga *et al.*, 2001) affect the divergence angle (*d*) in *Arabidopsis* (Appendix 1, Chapter 4).

Therefore, an accurate phyllotaxis model should be able to simulate both regular patterns and variations in patterns that occur in wild-type plants or in genetically modified plants.

Dynamical Systems and Irregular Patterns

Although theoretical phyllotaxis models are based on the regular arrangement of elements, can they also account for irregular or perturbed patterns? The FTOP predicts that divergence angles can vary within theoretical limits; however, what happens when these variations occur stochastically outside these limits? Are irregular sequences of divergence angles completely stochastic or can they include short regular sequences? Deterministic systems can accommodate a certain range of fluctuations. For example, in Douady and Couder's dynamical model (1996a,c), isolated spontaneous changes in the divergence angle (137°, 275°, 222.5°, 275°, 137°) can occur when the number of opposed parastichies (m and n) is high (e.g. (8, 13)) without disturbing or disrupting the global pattern. This phenomenon was observed on the stem of *Helianthus annuus* (Couder, 1998).

Certain theoretical models have well-established that the divergence angle oscillates randomly between certain intervals of decreasing amplitude towards a particular limit (Adler, 1974; Williams, 1975; Jean, 1994). This phenomenon was observed in many plants. For example, in *Arabidopsis*, the divergence angle stabilises at ~137° after the production of five leaves. Smith *et al.* (2006b) developed a deterministic model that reproduces these stochastic variations of the divergence angle in young plants (Fig. 3.2).

However, the irregular patterns found in many phyllotactic mutants result in new theoretical problems that cannot be solved using models developed for analysing regular phyllotactic patterns, because the extent of variation of divergence angles is

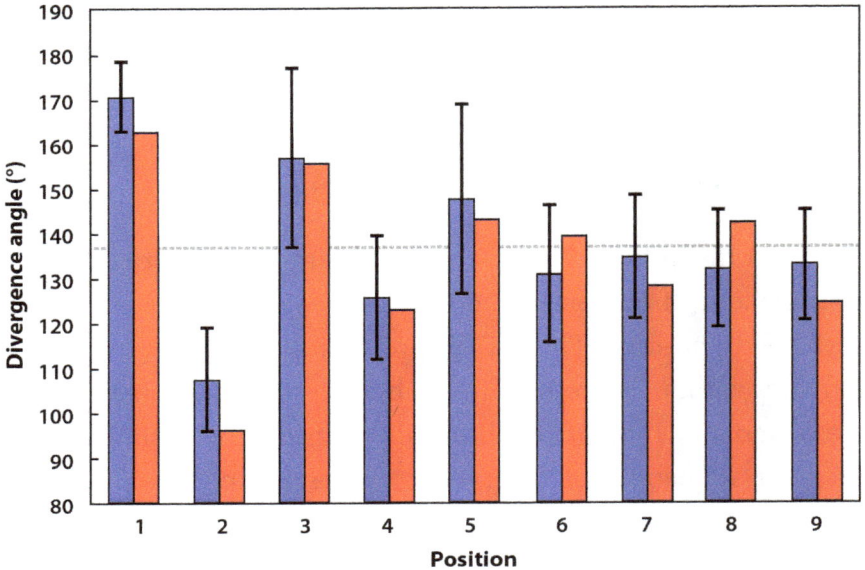

Fig. 3.2. Comparison of divergence angles: angles measured in *Arabidopsis* with standard error bars (blue) and angles generated by spiral phyllotaxis model (red). (From Smith *et al.*, 2006b; copyright [2006] National Academy of Sciences.)

too high. Most deterministic models are robust enough to consider a small amount of random variation (King *et al.*, 2004; Smith *et al.*, 2006b). What happens in terms of mechanistic processes when this variation is large enough to cause a switch from one type of phyllotaxy to another? Guédon *et al.* (2013) noted that dynamical systems would not be able to generate realistic phyllotactic sequences that include frequent and complex permutations[1] in the divergence angle. For irregular phyllotactic patterns, the main problem is the characterisation of their degree of order (Itoh *et al.*, 2000; Barabé and Jeune, 2004; Jeune and Barabé, 2004) and the identification of regular and irregular sequences of divergence angles (Guédon *et al.*, 2013). Perturbed patterns can be considered by using a probabilistic approach to phyllotaxis.

---

[1] Change in order of development of primordia.

Information Entropy Concepts Applied to Phyllotaxis

Shannon and Weaver (1949) noted that the average quantity of information per element in an overall set of elements (e.g. angles) is

$$H(x) = \sum_i H(x_i) = -\sum_i (p_i)\ln(p_i) \tag{1}$$

where (p_i) is the frequency of occurrence of individual elements x_i. This formula has been used to calculate the degree of order of a system in many biological disciplines (e.g. Atlan, 1972; Gatlin, 1972; Brooks and Wiley, 1986; Avery, 2003; Ricard, 2003).

Jean (1976, 1980) first introduced an entropy-like concept in phyllotaxis. Later, Marzec (1987) linked the concentration of a chemical substance to Shannon's information entropy concept. In Jean's model (1994), the bulk entropy (E_b) of a phyllotactic pattern is defined by

$$E_b = -\sum_{T=1}^{w} \log\big(S(T)/X(T)\big), \tag{2}$$

where T is the level in the hierarchy[2] and w, the rhythm of the hierarchy, that is, the order of the smallest growth matrix generating the hierarchy. $S(T)$ is the relative frequency of bifurcation

---

[2] Based on Church's idea (1904), Jean (1976, 1979, 1994) interpreted phyllotaxis as a hierarchical structure represented by a relational tree. It corresponds to a multilevel system. Each level contains a given number of nodes (corresponding to leaves or other appendicular organs) and is generated by the preceding levels. The time separating the generation of two levels corresponds to discrete theoretical units of time of any order.

The phyllotactic pattern represented in the relational tree (see figure below in footnote) corresponds to the Fibonacci series. Each level contains a number of nodes belonging to the Fibonacci series so that the number of nodes (F_n) from level 0 to level $n(T)$ corresponds to (1, 1, 2, 3, 5, 8, ...). If u_k is the number of nodes in level k, a hierarchy corresponding to the Fibonacci series has the form $u_0 = u_1 = 1$ and $u_{k+2} = u_{k+1} + u_k$. In the relational tree, a number of elements belonging to the Fibonacci series is added at each new level. Jean (double node) up to level T, and it represents the system stability. According to Jean (1994), the factor $-\log S(T)$ represents negentropy. $X(T)$, a parameter that increases rapidly with T, represents the pattern complexity; it is given by

$$X(T) = \prod_{k=1}^{T} k^{f(k)} \tag{3}$$

where $f(k)$, $k = 1, 2, 3,\ldots$, is the number of nodes in level k of the hierarchy. A detailed analysis of the results and predictions of

---

(1976, 1994) associated the growth represented by different trees to matrices linked to the generation of each level from the preceding ones.

The general rule is $F_{n+2} = F_{n+1} + F_n$, where F_n is the number of nodes in level n. Jean (1979) wrote that for a Fibonacci series where there is an injection of levels n and $(n+1)$ in level $(n+2)$ represented by the growth matrix

$$Q = \begin{bmatrix} 1 & 1 \\ 1 & 0 \end{bmatrix}, \qquad \text{we have } Q \begin{bmatrix} F_{n+1} \\ F_n \end{bmatrix} = \begin{bmatrix} F_{n+2} \\ F_{n+1} \end{bmatrix}.$$

A different matrix will correspond to a different type of ramification. For a Fibonacci tree, the order of the smallest growth matrix (Q) that generates the hierarchy is equal to 2. The growth matrix is discussed in detail in Jean (1976, 1994).

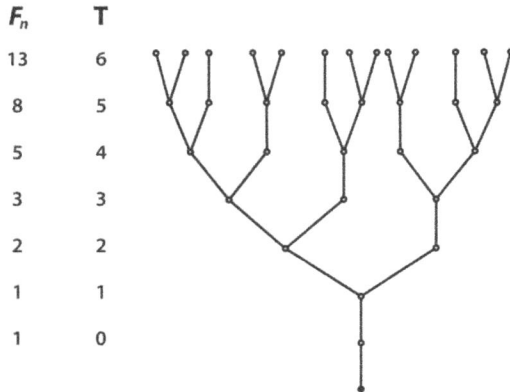

F_n	T
13	6
8	5
5	4
3	3
2	2
1	1
1	0

Relational tree of normal asymmetric phyllotaxis. Each generation (T) or level contains a Fibonacci number (F_n) of double or single nodes. (From Jean, 1976.)

Jean's entropy formula Eq. (2) is given in Jean (1994, Chapter 6). E_b is the production cost of a spiral pattern (m, n). Jean used this formula to find the hierarchy that minimises E_b for a given number of primordia and the various spiral phyllotactic patterns (m, n) that minimise the entropy production (function E_b), where (m, n) represents a visible parastichy pair. In this model (Jean, 1994; Jean and Barabé, 2001), the entropy for each set of visible parastichies (m, n) is given by

$$E_b = \log[m(m + n)/(n - m)n] + (m + n)\log 2. \qquad (4)$$

E_b increases with the number of parastichies. In this case, there is no maximum value, and E_b can theoretically be infinite for whorled systems. According to the model, whorls are special cases of spiral patterns; therefore, E_b can be calculated only if we know the spiral pattern it most likely represents. Jean and Barabé (2001) used this concept to calculate the bulk entropy (E_b) in the inflorescences of Aroids. However, it is difficult to use Jean's model to analyse the degree of order in disorganised systems because the entropy of phyllotactic systems cannot be determined without knowing the number of opposed parastichies present.

Marzec's theoretical model involves the concentration of a morphogenetic substance $p(\mu)$ produced by leaves and Shannon's informational entropy concept as given by

$$S = -2\pi \langle C \rangle \int_0^{2\pi} p(\mu) \ln p(\mu) du \qquad (5)$$

where S is the entropy and $\langle C \rangle$, the mean concentration. In this model, the divergence angles of 137.5°, 99.5°, 78.0°, 64.1°, 151.1°, 54.4°, 158.1°, and 106.4° that appear in various spiral phyllotactic systems are those corresponding to minimal entropy.

Practically, Shannon's entropy concept was used directly to calculate the degree of order of phyllotactic systems. For example, Vakarelov (1998) calculated the variation in entropy for *Pinus mugo* twigs by documenting the frequency of different phyllotactic patterns (m, n) in relation to altitude. He concluded that the priority order of spiral patterns established by Jean (1994) based on his entropy model was preserved. The degree of order (entropy) of a phyllotactic system, whether a regular or a random phyllotactic pattern, can be calculated using the entropy of a set of divergence angles. The minimum entropy corresponds to the greatest level of order in a phyllotactic system and is obtained when primordia are regularly distributed. Barabé and Jeune (2006) applied Shannon's entropy concept to *forever young* *(fey)* mutants described by Carlos *et al.* (1994). A set of phyllotactic angles was shown to correspond to a given value of Shannon's entropy. These values characterise whether a phyllotactic system is more or less organised and ordered *(fey* mutant 3) or random *(fey* mutant 1). Brillouin's (1959), Chaitin's (1966, 1987), Kolmogorov's (1965), and Yagil's (1985) concepts of entropy can also be used to describe the degree of order in perturbed phyllotactic systems. Barabé and Jeune (2006) applied these concepts to perturbed patterns in *fey* mutants and obtained similar results to Shannon's entropy. Itoh *et al.* (2000) observed that successive leaf primordia of *sho* mutants in rice were initiated at random positions on the SAM with a mean divergence angle of 110°. They used the entropy of a partition (Eq. (6)) to quantify the degree of order of phyllotactic patterns.

In the entropy of a partition based on Shannon's formula, the probability P_i (Eq. (6)) is estimated by the frequency of the number of elements n_i in a sector i (n_i/n), where n is the total number of elements. Then, in a centric representation with the apex being divided into i equal circular sectors ($i = 4$ in Itoh *et al.*, 2000), the entropy (E) is calculated as follows:

$$E = \sum_i \{E_i\} = -\sum_i \left\{ \frac{n_i}{n} \text{Log}_2 \left(\frac{n_i}{n} \right) \right\}, \tag{6}$$

where $E_i = 0$ if $n_i = 0$ in a centric representation.

Because the entropy of a partition cannot discriminate between an organised spiral system and a random phyllotactic system with the same number of leaves (Barabé and Jeune, 2004), a spiral system will appear positionally disorganised although it is a well-ordered pattern. Therefore, except for whorled and distichous patterns, the entropy of a partition cannot precisely represent the degree of order in a phyllotactic system, unlike the entropy calculated using a set of divergence angles (Barabé and Jeune, 2006).

However, although entropy represents the degree of complexity of a phyllotactic system, it does not provide a quantitative threshold for the transition between a regular and a random phyllotactic system. Jeune and Barabé (2004, 2006a, b) used a statistical approach to determine the threshold for transition from a regular to a random system.

Statistical Methods

The main problem faced in the analysis of phyllotactic mutants is determining the type of phyllotactic system and its degree of order. For example, are we dealing with a spiral, whorled, or stochastic phyllotactic system? How many randomly distributed elements can a phyllotactic system include without losing its regularity? These questions can be answered using a classical statistical approach that allows a global analysis of the order in phyllotactic systems (Jeune and Barabé, 2006a) or a combined approach using sequences of divergences angles (Guédon *et al.*, 2013).

A statistical approach based on the distribution of leaves in a planar representation (Fig. 3.1) was developed to determine the degree of order of phyllotactic systems. A chi-square (χ^2) test can be used to statistically differentiate between whorled and distichous patterns (aggregated dispersion), spiral patterns (uniform dispersion), and random patterns (random dispersion) (Jeune and Barabé, 2004, 2006a,b) (Fig. 3.3). The authors developed an equation (Eq. (7)) linking the empirical value of χ^2 to a measure of order in a phyllotactic pattern. This statistical test can be used to determine the number of regularly distributed leaves (spiral

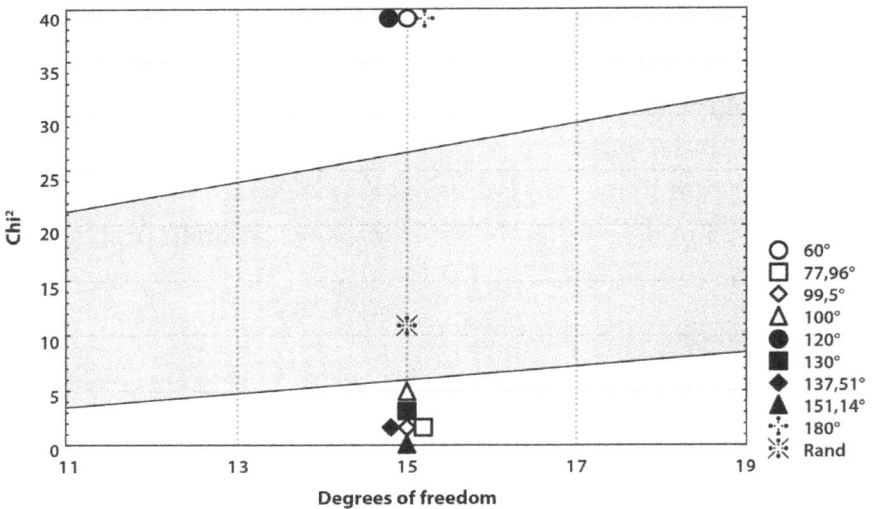

Fig. 3.3. χ^2 values for different phyllotactic systems: spiral (77.96°, 99.5°, 100°, 130°, 137.51°, and 151.14°), verticillate (60°, 120°, and 180°), distichous (180°), and random (Rand). The straight lines limit the confidence interval for random distributions; below this interval, distributions are regular (spiral), and above it, they are verticillate (or distichous). The degree of freedom is $(k - 1)$, where k is the number of theoretical sectors dividing the apex. (From Jeune and Barabé, 2006b; reprinted with permission from World Scientific Publishing.)

or verticillate) and random leaves in a typical phyllotactic pattern (distichous, spiral, or verticillate).

$$Z = E\left(\frac{SS_y}{y}\right) = (k-1)\frac{n-n_s}{n} - n_v\frac{k-1}{n} + n_v^2\frac{k-p}{np}. \qquad (7)$$

Eq. (7) corresponds to Jeune and Barabé's general formula (Eq. 4.11, 2006a) to calculate the number of leaves that are randomly or regularly distributed in a phyllotactic system. Z is the χ^2 value; $E\left(\frac{SS_y}{\bar{y}}\right)$ is the expectation of $\left(\frac{SS_y}{\bar{y}}\right)$, where SS_y is the sum of squares of deviates; $\bar{y}$ is the average number of leaves per sector; p is the number of orthostichies in the wild type; n is the total number of leaves; n_s is the number of leaves in a spiral pattern; n_v is the number of leaves in a whorl pattern; and k is the number of sectors dividing the SAM.

Based on this information, the authors performed numerical simulations with different numbers of leaves distributed randomly or regularly in the same phyllotactic system. By using this formula, the quantitative relationship between ordered and non-ordered elements in a phyllotactic system can be described in terms of the probability of appearance of a particular pattern. This can also be used to determine transitions from a disorganised system to an organised one by calculating the expectation of finding a particular phyllotactic pattern within a confidence interval (Jeune and Barabé, 2006a). This equation was modified by setting $n_s = 0$ and then used to estimate the number of leaves that are regularly distributed (n_v) in *sho* mutants (Table 3.1), where the wild type is characterised by a divergence angle of 180° (two orthostichies). The *sho 2* mutant (Itoh *et al.*, 2000) corresponds to a distichous system although it contains many randomly distributed leaves (Table 3.1, Jeune and Barabé, 2006b).

Table 3.1.   Phyllotactic variables calculated or estimated in *sho* mutants described by Itoh *et al.* (2000).

Mutants	*sho 1-1*	*sho 2*	*sho 3*
p	2	2	2
n	34	36	33
χ^2	16.82	32.44	18.88
Distribution	Random	Distichous	Random
$\hat{n}_v$ (estimated)	4.2	10.6	5.5
α (estimated)	96°	52°	80°

p — number of orthostichies; n — the total number of leaves; χ^2 — chi-square; $\hat{n}_v$ — number of distichous leaves; α — uncertainty in divergence angle (from Barabé and Jeune, 2006).

A second statistical test can be used to determine the variation in the divergence angle of a given leaf in relation to the preceding one (Jeune and Barabé, 2006b). By determining the range of uncertainty of the divergence angle, the point (leaf rank) at which a system becomes completely disordered can be determined. There is a quantitative link between the approach based on the degree of uncertainty of the divergence angle and that based on the distribution of regularly and randomly distributed leaves. Jeune and Barabé (2006b) developed a formula to quantify this relationship using numerical simulations involving different ranges of uncertainty of the divergence angle in three different phyllotactic patterns: distichous (two orthostichies), opposite-decussate (four orthostichies), and spiral (137°). A χ^2 test was performed to statistically determine the threshold of transition based on the uncertainty[3] in the divergence angle between different phyllotactic patterns.

---

[3] The uncertainty corresponds to the range of variation of the divergence angle.

Uncertainty in Divergence Angle

The following equation links χ^2 to the uncertainty (α) in the divergence angle.

$$\overline{\chi}^2 = a + be^{c\alpha} \qquad (8)$$

The parameter values were obtained by fitting the equation to numerical simulations by using the least-squares methods for distichous, opposite-decussate, and spiral patterns. In spiral systems, $a \approx -b \approx -1520$ and $c = 0.00008$. If the phyllotactic system is distichous, $a = k - 1$, $b = n\frac{k-p}{p} - (k-1)$, and $c = -0.05$ (Jeune and Barabé, 2006b).

Combining Eqs. (7) and (8) gives a quantitative relationship between the number of regular leaves (n_r) in a phyllotactic system and the uncertainty (α) in the divergence angle (Jeune and Barabé, 2006b).

This approach was applied to *sho* mutants (Itoh *et al.*, 2000) to determine the range of variation of the divergence angle for each mutant (Table 3.1). As predicted, the number of estimated distichous leaves is inversely proportional to the estimated range of the divergence angle. Although the *sho 2* mutant has a distichous system like the wild type, the divergence angle shows broader variation in range. Theoretical estimations can, therefore, be used to calculate the number of regularly distributed leaves and the range of uncertainty linked with the divergence angle in a partially random phyllotactic system. This type of statistical analysis can also reveal a pattern that is not visible from a planar representation of the SAM, as is the case in *sho* mutants where the range of uncertainty in the divergence angle is not visible using a schematic alignment of leaves.

For this statistical test to discriminate between regular and random patterns, more than 20 leaves $(n \geq 20)$ are needed and the SAM has to be divided into 16 sectors ($k = 16$). When these requirements are not met, it becomes more difficult to statistically distinguish among the main types of phyllotactic patterns (Jeune and Barabé, 2004). Additionally, although this approach is useful in characterising phyllotactic patterns, it does not discriminate between patterns within the same category (i.e. spiral systems).

Jeune and Barabé (2006b) showed that a spiral pattern consisting of 50 leaves appears to have a random distribution if ~20 leaves are randomly distributed. If the number of leaves in the spiral pattern increases to 80, ~33 randomly distributed leaves are needed for the pattern to appear random. Interestingly, this proportional ratio follows the Fibonacci sequence closely when n is a multiple of a number belonging to this sequence.

Although predictions can be made regarding the global variation of phyllotactic patterns, the statistical analysis of the distribution of leaves in a plane does not consider the biological mechanisms involved in the appearance of disorganised patterns and the interdependence of elements or the type of aperiodicity in a sequence of divergence angles along a stem. These approaches do not constitute developmental models of phyllotaxis based on biological processes, nor are they tools to analyse the type of order periodicity involved in sequences of divergences angles. They represent a global analysis of the degree of order in the SAM that allows us to determine the degree of randomness in a disorganised phyllotactic system.

Combinatorial Analysis of Irregular Patterns

Refahi *et al.* (2010, 2011) developed a combinatorial model and an algorithm for characterising regular and irregular patterns

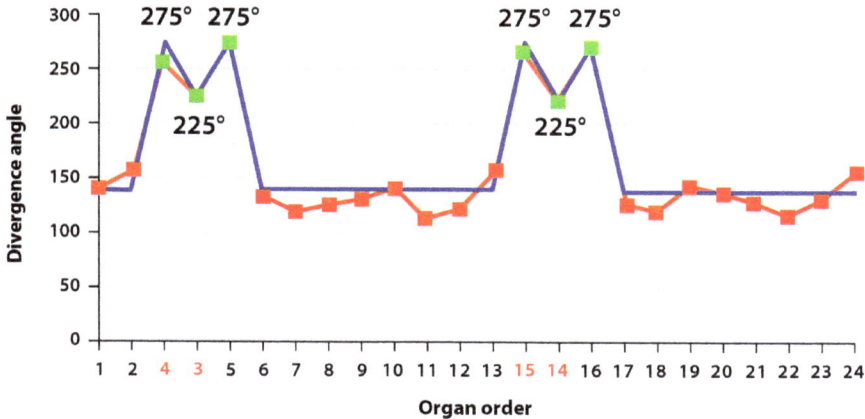

Fig. 3.4. Optimal labelling of a wild-type divergence angle sequence. The observed divergence angles are indicated by red squares for canonical angles[4] within baseline segments and by green squares for divergence angles within permuted segments (275°, 222.5°, 275°). The predicted divergence angle sequence is indicated in blue. The predicted organ order is given below (2-permutations in red). (From Guédon *et al.*, 2013; reprinted with permission from Elsevier.)

observed in wild-type and *Arabidopsis histidine phosphotransfer protein* 6 (*AHP6*) mutants of *Arabidopsis* (Mähönen *et al.*, 2006). These authors frequently observed an M-motif (275°, 222.5°, 275°) in successive divergence angles between flowers in the wild type (Figs. 3.4 and 3.5) and more often in mutants (Fig. 3.6). The phyllotactic sequences in wild-type plants are characterised by frequent occurrences of permutations[5] involving two flowers generally separated by a segment of canonical angles (137°) along the genetic spiral (Fig. 3.4). By contrast, *AHP6* mutants are characterised by permutations involving two or three flowers whose

---

[4] Typical angle of 137°.

[5] In a permutation involving two flowers (Fig. 3.5), flower #3, which normally forms an angle of 275° with flower #1, is positioned successively after flower #1. Then, flower #2 appears at an angle of 222.5° in relation to flower #1, between flowers #3 and #4.

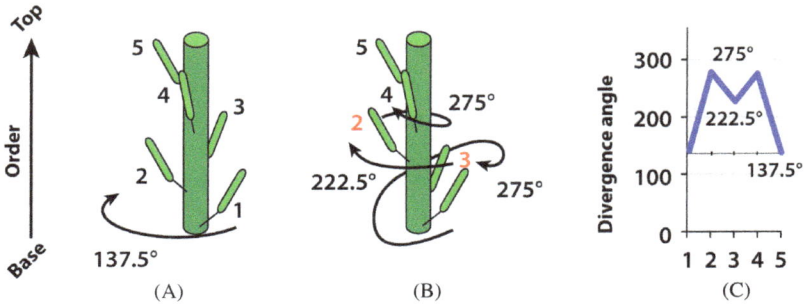

Fig. 3.5. Schematic representation of a 2-permutation: **(A)** normal succession of organs, **(B)** permutation of two consecutive organs, and **(C)** sequence of divergence angles corresponding to an isolated 2-permutation. (From Guédon *et al.*, 2013; reprinted with permission from Elsevier.)

succession corresponds to complex patterns (Fig. 3.4). By using empirical data from a mutant of *Arabidopsis*, Refahi *et al.* (2011) noted close agreement between the predicted and the measured divergence angle sequence.

Following the study of Refahi *et al.* (2011), Guédon *et al.* (2013) developed a probabilistic model to analyse complex patterns resulting from perturbations in the sequence of divergences angles in wild-type *Arabidopsis thaliana* and mutant *AHP6* exhibiting obvious defects in phyllotactic organisation. To label the sequences of measured divergence angles, they built three different models (stationary five-state hidden variable-order Markov chain, stationary seven-state hidden variable-order Markov chain, and seven-state combinatorial mixture model) and deduced a consensus from the three models. They showed that by using statistical and combinatory methods, a sequence that appears stochastic may contain similar perturbed segments. They also showed that the sequence of angles different from 137° can be explained by permutations in the order of insertion of two or three consecutive organs along the stem. Figures 3.5 and 3.6 show the optimal

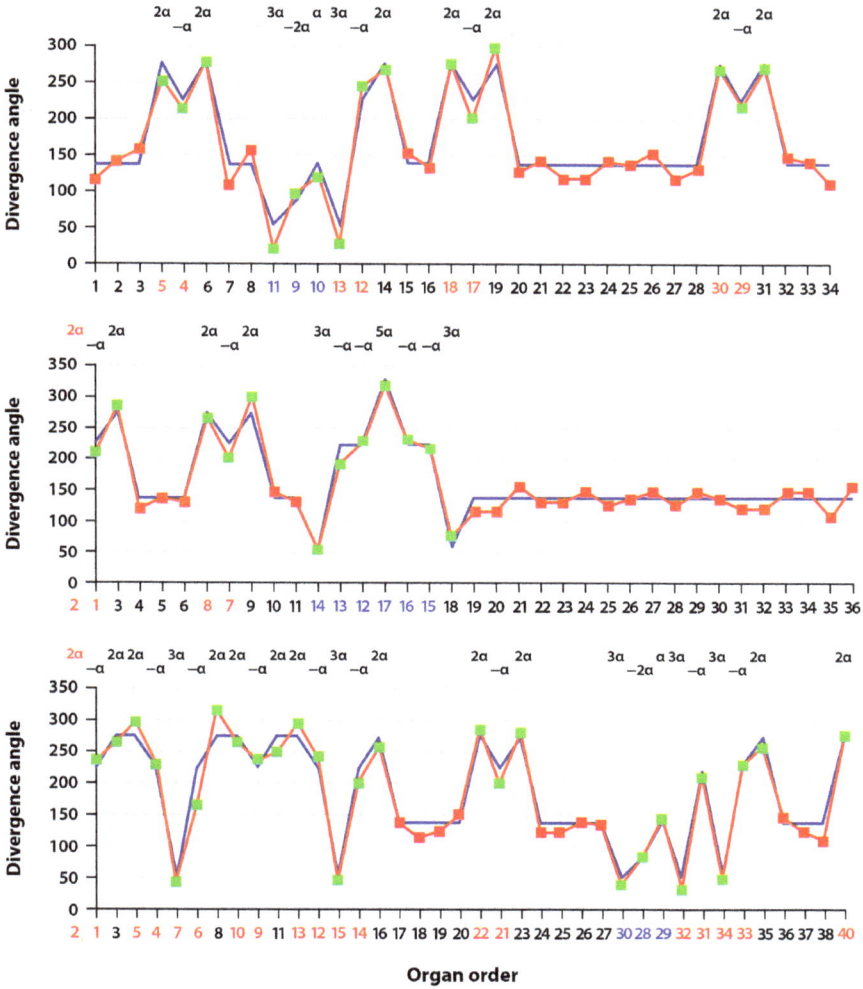

Fig. 3.6. Optimal labelling of three mutant (a, b, c) divergence angle sequences. The observed divergence angles are indicated by red squares for canonical angles within baseline segments and by green squares for divergence angles within permuted segments. The predicted divergence angles sequences are indicated in blue. The predicted organ orders are given below (2-permutations in red and 3-permutations in blue). The left-truncation of a permuted segment is hypothesised at the beginning of the last two sequences. (From Guédon *et al.*, 2013; reprinted with permission from Elsevier.)

labelling of a wild-type sequence and three mutant sequences in *Arabidopsis*.

In the wild-type, a segment of the sequence corresponding approximately to (275°, 225°, 275°) is observed frequently and more often in mutants. Guédon *et al.* (2013, p. 96) considered that this segment 'can be explained simply by a permutation in the order of insertion along the stem of two consecutive organs without changing their angular position' (Fig. 3.5).

In *Arabidopsis*, Guédon *et al.* (2013) empirically observed only the simple pattern $[2\alpha - \alpha 2\alpha]$ (275°, 225.5°, 275°) corresponding to an isolated 2-permutation. However, they theoretically inferred other patterns. For this purpose, they estimated hidden first-order Markov chains based on overall measured divergence angle sequences. A Markov chain is a stochastic model that describes a sequence of possible events in which the probability of each event depends on the state attained in the previous event. The states of these non-observable Markov chains represent theoretical angles. For each angle, they also calculated the circular standard deviation (von Mises distribution). They selected five divergence angles: 137.5°, 275°, 222.5°, 52.5°, and 85°. Based on these values, by using the variable-order Markov chain method, they identified the dependence between successive theoretical angles that could correspond to the most frequent permutation patterns in the empirical divergence sequences. Based on their analysis, they hypothesised that the segment with angles differing significantly from 137° could be explained by permutations involving two or three consecutive organs, for example, sequence order (3 2 1 4) (52.5°, 222.5°, 222.5°, 52.5°). Then, sequences that *a priori* look completely irregular can present stable repetitive segments such as (275°, 222.5°, 275°).

A sequence of divergence angles that looks completely stochastic may contain segments of regular permutations appearing at different positions in the sequence. Permutations along the generative spiral could partly explain the frequent occurrence of divergence angles that differ from 137° in Fibonacci phyllotactic systems. Based on their analysis, Guédon *et al.* (2013) concluded that the frequency and complexity of permutation patterns are genetically regulated and that 'permutation represents an intrinsic characteristic of phyllotaxis even in lower phyllotactic systems such as the one of *Arabidopsis*' (p. 104). which corresponds to a (3, 5) number of opposed parastichies.

Developmental Origin of Permutations

How do permutations occur? Guédon *et al.* (2013) formulated three simple hypotheses concerning the developmental origin of permutations:

Successive organs are initiated on the SAM with a typical divergence angle of 137° but in an inverted order of appearance. It would be interesting to find out if permutations are based on PIN1-mediated auxin transport.

Successive organs are initiated simultaneously with a typical divergence angle. Consequently, these organs are either positioned in the normal order or permutated as result of post-meristematic growth.

Successive organs are initiated with a typical divergence angle and follow the normal sequence. However, during early developmental stages, the younger organ could grow faster than the older organs. Then, post-organogenesis, the younger organ would connect to a lower level on the stem.

Importantly, the sequences in the model of Guédon *et al.* (2013) are not generated theoretically but are deduced from a statistical analysis of empirical sequences. This promising model could, therefore, be considered a theoretical-empirical model.

Does a biological explanation exist for the apparently random distribution of leaves? How can a change in one phyllotactic parameter lead to a disordered or perturbed phyllotactic system? Simulations with probabilistic approaches suggest that variations in the divergence angle and the size and form of the SAM are the main factors at play in the development of perturbed patterns (Jeune and Barabé, 2006b). Couder (1998) noted that the existing lattice determined by the value of the ratio of the diameter of leaf primordia to that of the SAM (van Iterson's parameter b) defines the position and formation of new primordia. In numerical simulations, Couder (1998) observed that disordered patterns arise when the simulation is initiated with a small ratio of the diameter of leaf primordia to that of the SAM. This corresponds to a large number of primordia or smaller primordia in relation to the SAM size. Disorganised phyllotactic patterns can also arise with a small number of primordia when there is some degree of uncertainty in the divergence angle of the first primordia that are initiated (Jeune and Barabé, 2006b). Additionally, phyllotactic mutants are often characterised by a variation in the SAM diameter (Clark *et al.*, 1993a,b; Jackson and Hake, 1999; Itoh *et al.*, 2000; Smith *et al.*, 2006b). This variation might theoretically explain the random fluctuation of the divergence angle that leads to perturbed phyllotactic patterns (Jeune and Barabé, 2006b). Recently, by using three different genotypes of *Arabidopsis*, Landrein *et al.* (2015) experimentally showed a positive correlation between the SAM size and the number of permutations in the sequence of divergence angles.

The ratio of the diameter of a primordium to that of the apex, in relation to the three-dimensional shape of the apex, constitutes an initial genetically based condition that constrains the elaboration of the phyllotactic system. In other words, it constitutes the boundary conditions of the dynamic phyllotactic system that generates the patterns (Marzec, 1999a).

Irregular Patterns and Noise in a Dynamical Model

Theoretical approaches that incorporate statistical elements in a dynamic model, such as the standard deviation of a Gaussian function (Koch *et al.*, 1998) or probability of occurrence of a primordium (Selvam, 1998), have rarely been reported. In their model of optimal light capture, King *et al.* (2004) introduced random fluctuations of the divergence angle in what is mainly a deterministic system. Karmakar and Key (2004) made a mathematical contribution to the probabilistic approach to phyllotaxis. They used Goodall's (1991) *cut-grow* map phyllotactic model, where the arrangement of leaves is determined by two parameters — λ *(cut)* and η *(grow)* — and demonstrated that when these parameters vary randomly, the system is not stable.[6]

Mirabet *et al.* (2012) addressed the problem of stochastic variability by using a dynamical system of interacting inhibitory fields. They showed that introducing noise in a dynamical model similar

---

[6] In Goodall's model, a newly formed leaf is first cut off along a line from the SAM; then, it grows rapidly and brings the adjacent tissue to iterate a new configuration. This process leads to a sequence of shapes constructed by a cut-growth map of triangles. Physically, λ (cut) represents the point where the edge between the meristematic surface creases and the newly initiated primordia cross the opposite edge of the triangle. The parameter η (growth) is the growth of the edge between the meristem and the newly initiated primordium by an extension factor $\eta \geq 1$ (Goodall, 1991).

to that developed by Douady and Couder (1996b) can result in the more frequent occurrence of permutations. They found three main types of defects in the initial spiral phyllotaxy: reversal of the handedness of spirals, concomitant initiation of organs, and occurrence of distichous angles.

They also introduced noise at different levels by modifying specific parameters: threshold E of the inhibitory field, necessary for the initiation of a new leaf (concentration noise),[7] and the radius R of the apex (size noise).[8] Both types of noise generate a random fluctuation of Γ, which corresponds to the ratio of the range of inhibition d to the radius of the apex (d/R) (Fig. 3.7).

The degree of variability of the divergence angle and plastochrone ratio is proportional to the strength of noise (Fig. 3.7). Threshold noise produces two main alterations. Mirabet *et al.* (2012) claimed that their numerical simulations reproduced sequences close to (137.5°, 275°, 222.5°, 275°, 137.5°), which represent M-shaped sequences of angles which were also observed in *Helianthus annuus* (Couder, 1998) and *Arabidopsis* (Refahi *et al.*, 2011; Fig. 3.5). Threshold noise also results in the appearance of divergence angles close to 180°, corresponding to a transient distichous phyllotaxis. This type of alteration was reported for some *Plethora mutants* (*plt3, plt5, plt7*) of *Arabidopsis* (Prasad *et al.*, 2011).

For apex size noise (R), as for threshold noise, sequences close to (ϕ, 2ϕ, $360 - \phi$, 2ϕ, ϕ) and distichous 180° angles also occur, but less frequently. In contrast with threshold noise, the handedness of the spiral is occasionally reversed. Based on their numerical simulations, Mirabet *et al.* (2012) concluded that rapid

---

[7] Before each new initiation, the threshold is defined randomly according to a Gaussian distribution.

[8] Following each new initiation, the radius is redefined randomly following a Gaussian distribution.

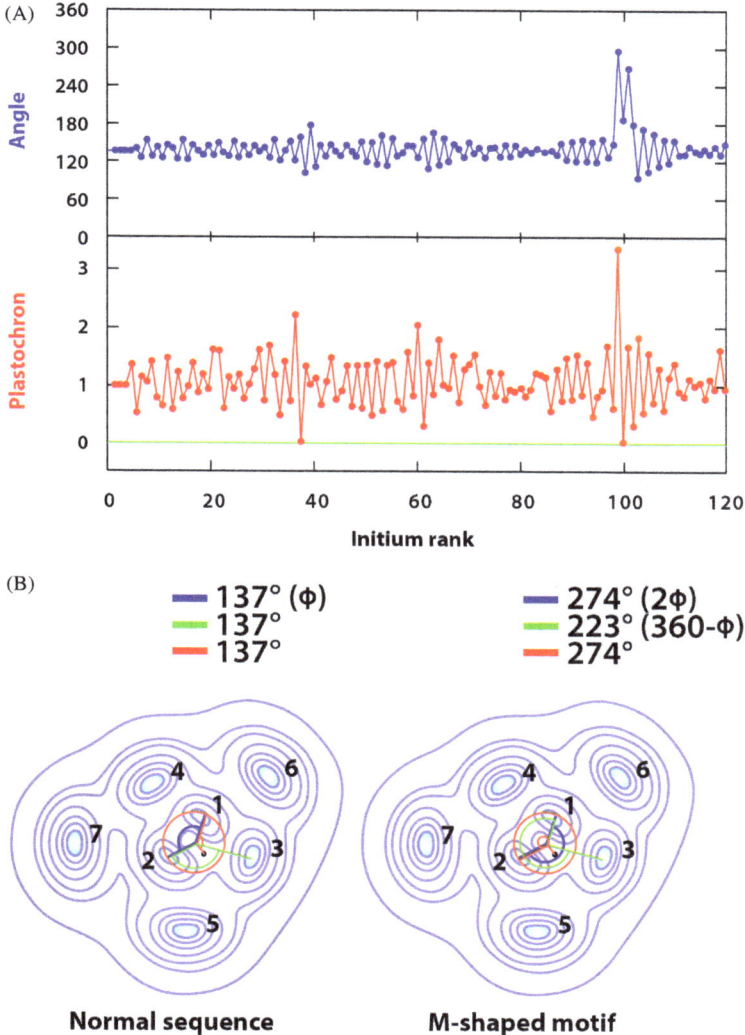

Fig. 3.7. The model with noise on the threshold for organ initiation. **(A)** A typical sequence of angles and plastochrones (time delay between consecutive initiations); the simulation is started with no noise; M-shaped angle sequences correspond to concomitant initiations (vanishing plastochrone). **(B)** Schematic explaining the origin of M-shaped sequences. A vanishing plastochrone implies two equivalent initia, which are ranked randomly. This either yields a sequence of divergence angles close to the golden angle $\phi \sim 137°$ or an M-shaped sequence of the type (137.5°, 275°, 225.5°, 275°, 137.5°). Simulation parameters: steepness of inhibition gradient $\alpha \sim 2$, $\Gamma \sim 1.92$, mean threshold for initiation $E = 1$, the standard deviation of threshold $\sigma_E \sim 0.8$. (From Mirabet *et al.*, 2012.)

fluctuations of Γ in deterministic models introduce noise in the system. They also proposed that threshold noise is more realistic because it reproduces sequences of angles observed *in vivo* in *H. annuus* (Couder, 1998) and *Arabidopsis* (Refahi *et al.*, 2011). They determined that threshold noise, which corresponds to signal noise, is the main source of stochasticity in the *Arabidopsis* apex.

Secondary Inhibitory Fields

Mirabet *et al.* (2012) considered that studies on auxin signalling (Vernoux *et al.*, 2011), hormones (Giulini *et al.*, 2004), and mechanical forces (Newell *et al.*, 2008a; Hamant *et al.*, 2008) indicated that other types of fields could play a role in the emergence of phyllotactic patterns. They used their phyllotactic model to also investigate the role of two secondary fields in the phyllotactic organisation: during and after the initiation of primordia. After a delay, each new primordium becomes the source of a second inhibition field of range d_2 that will interact at the same level as the first inhibitory field. The new initium will be formed at the point where the interaction between the two fields reaches the threshold E. This was shown to generate many stable phyllotactic patterns differing from the Fibonacci sequence. The effect of the second field was equivalent to changing the size of the apex permanently in the first field. In the second type (post-initiation correction), the physiological age of the primordium changes after its initiation, making it 'older' or 'younger'. They noted that in this case, the obtained sequences and histograms of angles are similar to those of the threshold noise model with no secondary field. Further, the number of alterations seems to decrease if the initium is 'younger' when it experiences the effects of a secondary field. They concluded that a secondary field acting on the state of development of organs can significantly improve the regularity of a noisy phyllotactic model.

Mirabet *et al.* (2012) considered that four different sources of variability occurring on different biological structures can explain disorganised patterns: (1) a stochastic variation in the position of the initium on the apex; (2) a stochastic variation in the auxin concentration; (3) a variable response of cells to the auxin signal; and (4) a stochastic variation of the diameter of the apex.

The model of Mirabet *et al.* (2012) provides a theoretical framework for analysing the nature of noise involved in irregular phyllotactic patterns. However, it contains many parameters that are difficult to link with biological processes.

Stochastic Manifestation of Phyllotaxis at Cellular Level

Among probabilistic approaches to phyllotaxis, Refahi *et al.* (2016) developed an original stochastic model based on inhibitory signals that integrate processes occurring at microscopic and macroscopic levels. In deterministic energetic models such as those of Douady and Couder (1996a–c), patterns of inhibitory fields determine the position of a new primordium on the SAM. In their computational model, Refahi *et al.* (2016) showed that random fluctuations in the way cells perceive these inhibitory signals can disturb phyllotactic patterns owing to a change in the timing of organ initiation. This model is explained further below.

Their model is based on Douady and Couder's classical deterministic model (1996b) relative to the stability of the inhibitory field generated by a primordium (see Chapter 2). In this model, a new primordium will be initiated where the inhibitory field representing the sum of the contributions of pre-existing primordia is under a predefined threshold. To analyse the effect of noise in the perception of the inhibitory field by the cells at the initiation site of a new primordium, Refahi *et al.* (2016) induced a perturbation in a stationary spiral pattern by forcing the initiation of a primordium at the second local minimum site instead of the global one. After implementing this theoretical modification, they noted that the noise in the perception of the order of initiation of primordia does not propagate far in the sequence of divergence angles and that the specification of angles in the classical (typical) system is strongly resistant to perturbations in the order of initiation of local minima. However, the plastochrone itself is affected over a much longer time span. These results indicate that the variation in the timing of organ initiation could be due to the combined effect of noise in the perception of inhibitory fields by the founder cells of the developing primordia and by the separation of spatial and temporal components of the self-organising-system.

Refahi *et al.* (2016) noted that at any time t, k cells located at the periphery of the central zone may potentially initiate the formation of primordia depending on the local value of the inhibitory field, which is represented by a signal in each cell. They assumed that the perception of this inhibitory field or signal by the founder cell (k) is stochastic. Each founder cell perceives the value of the local inhibitory field, $E_k(t)$, and commits to forming a primordium depending on the level of inhibition $E_k(t)$ it experiences and the duration δt for which it is exposed to this level of signalling. This is described by the following probability function:

$$P(X_k(t, \delta t) = 1) = \lambda(E_k(t))\delta t, \qquad (9)$$

where $X_k(t, \delta t$ represents the number of primordia initiated at cell k in the time interval $[t, t + \delta t]$; it takes a value of 0 or 1. λ, which depends on the local inhibition value $E_k(t)$ at cell k is interpreted as a temporal density of initiation. The authors noted that the influence of the inhibitory fields on the probability of initiation is expressed in the context of a number of constraints that frame the dependence of λ on local inhibition. Then, the rate parameter is expressed as (see Refahi *et al.*, 2016 for a detailed derivation of the function)

$$\lambda\left(E_k(t)\right) = e^{-\beta\left(E_k(t)-E^*\right)},\tag{10}$$

where E^* defines the sensitivity of the system to inhibition and β is the ability of the system to discriminate between inhibitions levels, that is, its capacity to react differently to close inhibition levels. Based on this equation, Refahi *et al.* (2016) deduced that for each cell k, the probability of initiating a primordium during time interval δt is expressed as

$$P(X_k(t,\delta t) = 1) = e^{-\beta\left(E_k(t)-E^*\right)}\delta t.\tag{11}$$

Interestingly, Refahi *et al.* (2016) proposed a stochastic formulation of their model at the cellular and macroscopic levels.

The probability that a primordium will be initiated is a non-homogeneous process regulated by the level of the inhibitory field at that site. At the cellular level, the recruitment of individual cells for organ initiation is stochastically independent for each cell. The expected[9] number of cells, $X(t, \delta t)$, independently recruited for primordium initiation during period δt can be estimated as $X(t,\delta t) = \Sigma_{k=1}^{K} X_k(t,\delta t)$. This expectation corresponds to the sum of the expectations of the individual independent Poisson processes:

$$E(X(t,\delta t)) = \sum_{k=1}^{K} X_k(t,\delta t) = \sum_{k=1}^{K} P(X_k(t,\delta t) = 1)$$

$$E(X(t,\delta t)) = \delta t \sum_{k=1}^{K} e^{-\beta\left(E_k(t)-E^*\right)}.\tag{12}$$

---

[9] In probability, the expected value of a random variable is intuitively the value that is anticipated on average when an experiment is replicated several times.

This equation gives an estimate of the expected number of peripheral cells that may independently be involved in primordium initiation during time δt in relation to the local inhibition field $E_k(t)$.

At the macroscopic level, a 'valley' of local inhibition can encompass several cells. Several founder cells can contribute to primordium initiation during period ∂t, thereby increasing the probability of primordium initiation in that valley. The stochastic processes of all cells k covered by a local inhibition valley l add up and together define a new stochastic process N_l responsible for the initiation of a new primordium. This higher-level process is also a Poisson process corresponding to the sum of independent Poisson processes with intensity $\Lambda_l = \Sigma_{k \in K_l} \lambda_k(E_k(t))$, where k represents each cell in the set K_l spanned by the valley of the lth local minima. Based on this equation, the expected number of primordia initiated during a small period ∂t is expressed as

$$E(N_l(t, \partial t)) = \partial t \sum_{(l=1)}^{L} \Lambda_l(t), \tag{13}$$

where L is the number of local minima.

As indicated previously, the preceding equations involved processes acting at two levels. At the microscopic level, there is a deterministic component related to the geometry and the dynamic of the fields $(E_k(t))$ and a stochastic component related to the perception of the inhibitory fields by cells (Eq. 10). At the macroscopic level, in each inhibition valley, more than one founder cell may contribute to primordium initiation. It follows that the probability of triggering primordium initiation increases with the size of the valley.

Interplay of Phyllotactic Parameters

How can we link the parameters of classical systems to those of this new stochastic model? The stochastic model of Rafahi *et al.* (2016) consists of two control parameters linking phyllotactic variables to the geometry and theoretical biological processes acting at the cellular level in the SAM: one linked to the number of permutations of divergence angles and the other, to the phyllotactic angle and plastochrone.

In the model of Refahi *et al.* (2016), the variation of the number of 2- and 3-leaf order permutations is a function of Γ (ratio of the diameter of the primordium to the radius of the SAM), β (ability of the system to discriminate between inhibition levels), E^* (sensitivity of the system to inhibition), and s (equivalent to parameter α, the force of the interaction or 'stiffness' as in Douady and Couder's model [1996a]). For $s = 3$, as in Douady and Couder (1996b), they empirically obtained a quantitative relation between the global number of perturbations and a combination of Γ, β, and E^* as $\Gamma_p = \Gamma\beta E^*$. This quantitative relationship indicates that the total number of perturbations in a phyllotactic sequence depends on both the SAM geometry (Γ) and the signal perception by cells (βE^*). This new parameter is considered a control parameter for perturbations.

Refahi *et al.* (2016) empirically showed that Γ, which is used in certain deterministic models, is not adequate as a control parameter in their stochastic model owing to the influence of β or E^*. On a particular phyllotactic branch of the deterministic phyllotactic tree (e.g. the Fibonacci branch), variations in β or E^* can have a significant effect on the divergence angle or plastochrone for a specific Γ value. For a given Γ value, more than one α value can exist. By using different combinations of parameters, they found that for $\Gamma_D = \Gamma^{2/3} \frac{1}{\beta^{1/6} E^{*1/2}}$, a unique value of the new control parameter can correspond to a quasi-unique value of divergence angle and plastochrone. As for the number of perturbations, the divergence angle and plastochrone are controlled by a combination of geometric (Γ, s) and perception parameters (β, E^*). The total number of parastichies is also inversely proportional to Γ_D, as is the case for Γ in deterministic models.

The appearance of phyllotactic patterns in the stochastic model of Refahi *et al.* (2016) results from a reaction to inhibitory fields at the level of individual cells and also the overall set of cells. The uniqueness of this model in relation to others like the stochastic model, where an initiation event is triggered by an inhibition threshold, is its focus at the level of individual cells. The probability that an initiation event will be triggered depends on the inhibition level in these individual cells. Refahi *et al.* (2016) considered that this provides a more realistic abstraction of the underlying signals and associated mechanism (Fig. 3.8A).

This stochastic model enables predicting the existence of frequent 2- and 3-leaf order sequence permutations and particular series of permutations (e.g. 4- and 5-permutations). Therefore, the disorganised phyllotactic patterns observed in mutants can be analysed in terms of control parameters by integrating the size of primordia, inhibitory fields, apex, and signal perception by individual cells. Refahi *et al.* (2016) observed that the number of perturbed sequences is inversely proportional to the plastochrone length. This can be explained by the fact that shorter plastochrones are linked to the development of a greater number of primordia on the SAM, thereby increasing the possibility of reversions in the order of appearance of primordia (Landrein *et al.*, 2015) and variations in the divergence angle

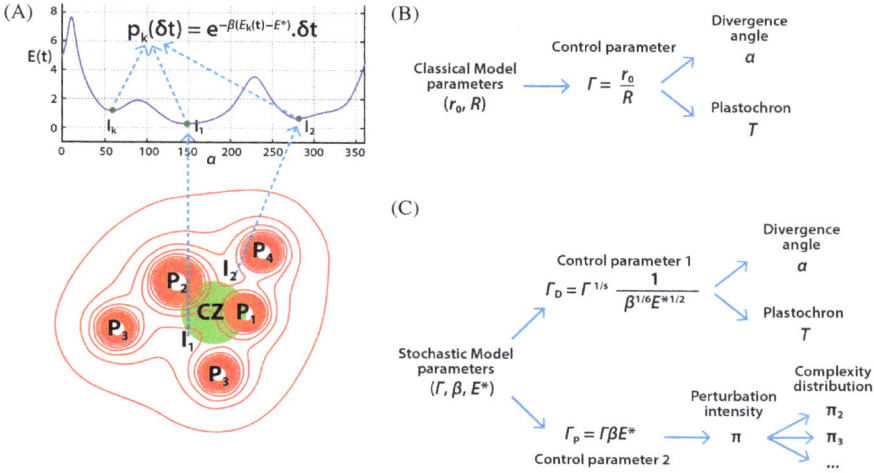

Fig. 3.8. Structure of the stochastic model. **(A)** Inhibitory fields (red), possibly resulting from a combination of molecular processes, are generated by primordia. In the peripheral region of the central zone (CZ, green), they exert an inhibition intensity $E(t)$ that depends on the azimuthal angle α (blue curve). At any time t and at each intensity minimum of this curve, a primordium can be initiated during a period δt with probability $p_k(\delta t)$ that depends on the inhibition level at this position. **(B)** Relationship between the classical model parameters and its observable variables. A single parameter Γ controls both the divergence angle and the plastochrone (T). **(C)** Relationship between the stochastic model parameters and its observable variables. The stochastic phyllotaxis model is defined by three parameters: Γ, β, and E^*. The observable variables α and T and π, π_2, π_3, … are controlled by two distinct combinations of these parameters: $\Gamma_D = \Gamma^{5/3} \frac{1}{\beta^{1/6} E^{*1/2}}$ controls the divergence angle and plastochrone and $\Gamma_p = \Gamma \beta E^*$ controls the global percentage of permuted organs (π), which in turn controls the distribution of permutation complexities: π_2, π_3. (From Refahi *et al.*, 2016.)

(Couder, 1998; Rutishauser, 1998). Stochasticity induces modifications in the plastochrone that correspond to the timing of primordium initiation.

The stochastic model of Refahi *et al.* (2016) can reproduce typical spiral and whorled patterns. However, theoretical results show that the noise affecting the plastochrone can generate disorganised phyllotactic patterns. More generally, this model highlights the relationship between global information, corresponding to the distribution of a signal, and local interpretation of this information by individual cells.

Conclusion

Until recently, few studies have elaborated a probabilistic phyllotaxis model; by contrast, many deterministic approaches to phyllotaxis are available. Most studies explaining the predominance of Fibonacci series in phyllotaxis still focus on a deterministic viewpoint. When can we expect a probabilistic analysis of the predominance of Fibonacci series in plants? The elaboration of a general probabilistic phyllotaxis model represents a new challenge or need for a higher level of explanation for both theoretical and experimental research.

4 Role of Genes in the Framework of Biochemical and Molecular Models

The dynamics of the appearance of primordia in a regular manner is determined by a combination of genetic, physiological, and physical factors. Different approaches to explain phyllotaxis have been developed over the last century. They range from physiological models to chemical and physical models (for a historical review of these models see Rutishauser, 1981; Jean, 1994; Adler *et al.*, 1997; Jean and Barabé, 1998b; Reinhardt *et al.*, 2000; Reinhardt, 2005a; Kuhlemeier, 2007; Pennybacker *et al.*, 2015). Biological explanations of phyllotaxis are closely related to models of shoot apical meristem (SAM) organisation and function in terms of patterns of cellular zonation as well as determinate/indeterminate developmental pathways (Lyndon, 1998; Tooke and Battey, 2003; Nakayama *et al.*, 2012; Guenot *et al.*, 2014).

The mechanisms involved in the regulation of phyllotaxis have been extensively studied through surgical experiments (Snow and Snow, 1931; Loiseau, 1969; Steeves and Sussex, 1989; Reinhardt *et al.*, 2005) or through the application of growth regulators (Wardlaw, 1949a,b; Snow and Snow, 1962; Schwabe, 1971; Charlton, 1974; Meicenheimer, 1981). The involvement of a chemical substance in the formation of phyllotactic patterns was first proposed by Schoute (1913), who hypothesised that a new developing primordium would produce a diffusible chemical substance that inhibited the initiation of other primordia in its immediate neighbourhood. This theory was further elaborated on by Richards (1948) and Wardlaw (1949a,b) based on empirical data. Later, Turing (1952) introduced the use of dynamical systems to model different biological phenomena, including phyllotaxis. However, the development of the theoretical models based on the diffusion of one or two substances (inhibitor and activator) has occurred mainly from the seventies onwards (Hellendoorn and Lindenmayer, 1974; Thornley, 1975; Mitchison, 1977; Veen and Lindenmayer, 1977; Young, 1978; Meinhardt, 1982; Schwabe and Clewer, 1984; Chapman and Perry, 1987; Yotsumoto, 1993; Meinhardt, 1998, 2003; Smith *et al.*, 2006a).

This chapter reviews information on pattern formation at the level of the SAM as well as mutations that result in abnormal phyllotactic patterns. A description of the most recent models and mechanisms involved in the formation of phyllotactic patterns is presented at the end.

Positional Information and Cell Lineage

Developmental studies have shown that positional information is critical for controlling the identity of cells and tissues (Dawe and Freeling, 1991; Scheres, 2001; Smith *et al.*, 2006a,b). Phyllotaxis provides a great system for studying positional information in plants. The position of a primordium may be determined by cell-to-cell signalling, diffusing signals (morphogens), mechanical forces, or other yet unknown mechanisms.

Pattern formation and signalling pathways are two different perspectives for examining morphogenesis. In the framework of pattern formation, it is assumed that global processes at the organ level, such as morphogenetic fields or mechanical forces (Mirabet *et al.*, 2012; Nakayama *et al.*, 2012; Guenot *et al.*, 2014; Pennybacker *et al.*,

2015), induce particular patterns. The differentiation pathway followed by cells is controlled by the processes acting at the level of the organ. In this context, Day and Lawrence (2000, p. 2985) argued that *'shape and size in both animals and plants is controlled in part by mechanisms that read absolute dimensions rather than cell number'.* On the other hand, in the framework of signalling pathways, cell-to-cell signalling can induce the formation of particular patterns. In this case, the patterns result from interactions controlled at the level of the cell. This reiterates the fact that a variety of mechanisms operate at the different hierarchical levels in an organism (i.e. from cell to the whole organ).

The function or overall necessity of cell division in plant development has long been debated (e.g. Kaplan, 1992; Doonan, 2000; Wyrzykowska *et al.,* 2002). Changes in the processes involved in cell division can alter the plant growth rate without affecting the global morphology of its shoot (Smith *et al.,* 1996; Cockcroft *et al.,* 2000; Wyrzykowska *et al.,* 2002). It has also been demonstrated that plant morphogenesis can occur via a cell division-independent pathway (Pien *et al.,* 2001; Wyrzykowska *et al.,* 2002). For example, Smith *et al.* (1996), described a recessive mutation of maize, *tangled-1 (tan-1),* that altered the orientation of cell division throughout leaf development without altering the overall shape of the leaf. Recently, Fleming's group showed experimentally that the modulation of cell wall extensibility is an alternative mechanism to cell division during plant morphogenesis, particularly in the organisation of the SAM (Pien *et al.,* 2001; Wyrzykowska *et al.,* 2002).

There appears to be no direct relationship between the pattern of stem-cell lineage in the SAM and leaf position (Poething, 1987; Wyrzykowska *et al.,* 2002). For example, changes in the cell division patterns through the overexpression of *Nicta;CycA3;2* and *Spcdc25* genes in the apical meristem of tobacco had no influence on meristem shape and phyllotaxis (Wyrzykowska *et al.*, 2002). Similarly, in Arabidopsis, Furner and Pumbrey (1992) found that the mechanisms that generate phyllotaxy and the initiation of axillary meristems occur in a position-dependant, lineage-independent mode.

Klar (2002) proposed that asymmetric stem-cell patterns of cell division offer a possible explanation for the *de novo* generation of Fibonacci patterns in phyllotaxis. This concept is based on the premise of an asymmetric cell division that originates from a mature cell that can divide further, and a juvenile cell that requires a period of 'growth' before it starts dividing. This hypothesis was rejected by Fleming (2002) who considered that in Angiosperms (p. 723) '*Most evidence supports short-distance chemical signalling as the mechanism involved in designating the site of primordium initiation, with biophysical alteration in cell-wall extensibility as a key executor of the morphogenetic programme initiated*'. Based on current studies related to chemical signalling mechanisms in plants, the latter view seems to be predominant.

Position of Primordia

Based on a morphological study of *Hedychium* (Zingiberaceae), Kirchoff (2000, 2003) suggested that two factors influence the position of floral organs: the position of the other floral organs present on the apex and the shape of the apex itself. At the physiological level, Reinhardt and Kuhlemeier (2002) proposed (p. 196) '*... that dynamic gradients of auxin in the meristem determine the site of organ formation, and that pre-existing leaves influence these auxin gradients by modulating acropetal auxin transport or auxin distribution within the meristem*'. Experiments using an inhibitor of polar auxin transport have shown that the pre-existing leaves do provide spatial information to determine the position of new primordia (Reinhardt *et al.*, 2000; Smith *et al.*, 2006a). The fundamental role of auxin in phyllotaxis will be covered in detail in subsequent sections of this chapter.

Green (1992a,b) proposed an alternative to the chemical model. According to this author, the site of organ initiation is determined by biophysical forces on the surface of the apex. The new primordium appears in an area of the SAM where the surface tension is at its minimum. The preceding primordia also act as border conditions for the positioning of newly formed primordia. More recently, Shipman and Newell (2004) demonstrated how phyllotaxis may be understood as the energy-minimising buckling pattern of a compressed shell (the plant's tunica) on an elastic foundation. Hamant *et al.* (2008) also explored the role of mechanical signals in developmental patterning (see Chapter 5). In that context, Selker and Lyndon (1996) have shown that the pattern of new leaf positioning in explants of *Nasturtium officinale* R.Br. depends on the existing boundary structures and on the newly formed primordia. Recent studies on the protein expansin support the interpretation of Green (1992a,b). However, the activity of the expansin gene could also support phyllotactic theories involving chemical signals (see above and Reinhardt *et al.*, 1998; Kierzkowski *et al.*, 2012). At the molecular level, it has been shown that the gene *SHOOT MERISTEMLESS (STM)*, involved in the integrity and maintenance of the SAM, responds to phyllotactic pattern information originated in pre-existing primordia (Long and Barton, 2000). Other genes such as *PIN1* and *PID* are induced at the site of organ initiation and are required for its formation (Hardtke and Berleth, 1998; Christenson *et al.*, 2000; Vernoux *et al.*, 2000; Reinhardt, 2005a; Kuhlemeier, 2007).

Recent studies exploring different potential mechanisms linked to pattern formation have highlighted the participation of several players at the molecular, physiological, and structural levels. These different aspects will be discussed in the following sections and in other chapters.

Genes and Their Effects on Pattern Formation

The most recent models considering molecular biology information are based on the principle of localisation by actively transported substances. The discovery of genes that are associated with the establishment of phyllotactic patterns could certainly represent a breakthrough in enhancing our understanding of phyllotaxis. Mutants of Arabidopsis and *Antirrhinum*, in particular, constitute important materials to study phyllotactic changes that occur during the growth and development of the stem. Similarly, the quantitative analysis of phyllotactic parameters in Arabidopsis mutants can provide new insights into the relationships between these parameters.

Recent molecular analyses have demonstrated that various genes regulate meristem development and leaf formation and occupy specific domains (Fig. 4.1) (Braybrook and Kuhlemeier, 2010; Sassi and Vernoux, 2013). According to Reinhardt *et al.* (2000), some of these genes encode transcription factors of the homeobox class, whereas others appear to be involved in cell-to-cell signalling. Considering the numerous modes of action of genes, it is logical and easy to assume that signalling pathways and pattern formation are contrasting concepts (Reinhardt *et al.*, 1998). However, the linkage between them may be more intimate than previously thought (Cummings and Strickland, 1998).

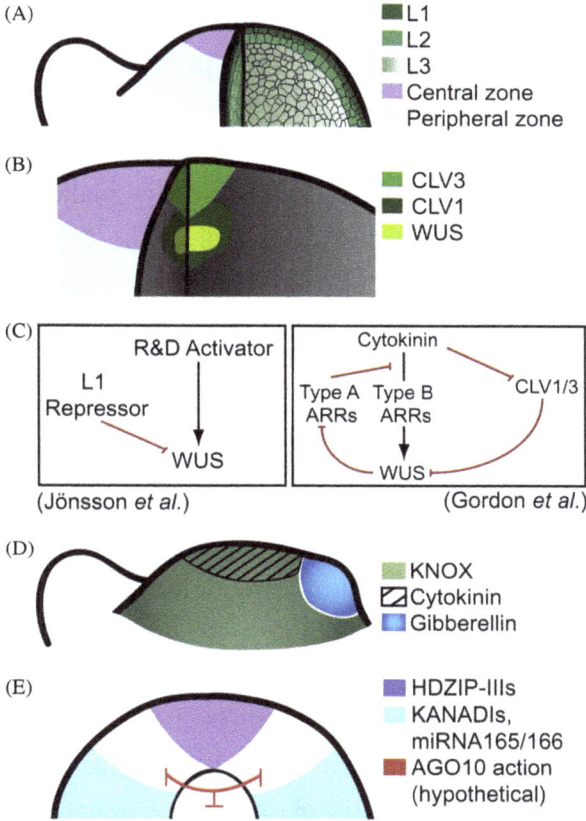

Fig. 4.1. The shoot apical meristem: a source for new leaves. (A) The organisation of the meristem in terms of functional zones and cell layers. (B) Expression domains of the basic *WUS/CLV* regulatory loop components. (C) Role of cytokinin and putative *R* (reaction) and *D* (diffusion) in the spatial regulation of *WUS*, as described by Jönsson *et al.* (2006) and Gordon *et al.* (2009). Red and black lines illustrate negative and positive interactions, respectively. ARRs, authentic responses regulators. (D) *KNOX* expression within the meristem and areas of high cytokinin and gibberellin (relative to each other). (E) Interaction between *HD-ZIPIIIs* and *KANs* with respect to meristem maintenance. HD-ZPIII expression demarcates the top of the peripheral zone and *KAN* expression the lower boundary. A diagrammatic representation of hypothetical ways for *ZWILLE/PIN-HEAD/AGO10* (red lines) to affect this boundary. (From Braybrook and Kuhlemeier, 2010; reprinted with permission from American Society of Plant Phisologists.)

In this context, we are confronted with questions such as: What is the exact role of genes in the production of phyllotactic patterns? Which physiological processes are involved? What is the role of plant growth regulators and the impact of physical forces on the dynamics of the SAM? Is cell-to-cell signalling the driving force, or is there a more global control at the organ level? The observed transitions between the different phyllotactic patterns that appear naturally or through surgical experiments, application of phytohormones, or radiation treatments suggest that genes are not the only players involved, and therefore the generation of phyllotactic patterns in plants is a complex series of mechanisms operating at multiple levels, from cells to organs.

Given that the four fundamental phyllotactic parameters (i.e. angle of divergence, number of parastichies, plastochrone ratio, and the angle between parastichies) are linked in the same formula (see chapter on "Phyllotactic Parameters"), a modification in one parameter will likely affect the values of the other three. Theoretical models predict that if the ratio between primordium size and meristem size (or the plastochrone ratio) decreases, the total number of parastichies increases (Jean, 1994). The same negative correlation exists between the size of the meristem and the leaf arc formed by the base of the leaf (Rutishauser, 1998). The empirical correlations between phyllotactic parameters are illustrated in the study of Rutishauser (1998) who conducted a survey of different phyllotactic systems in vascular plants. However, a fundamental question remains: Which parameter has a biological significance? In other words, among the four parameters commonly used in mathematical models, which one(s) is(are) directly under the control of a biological process in the developing apex? An explicit link between molecular biology and theoretical models remains difficult to demonstrate. In many mutants, the effects on phyllotaxis are indirect. Mutations modifying phyllotaxis usually affect two or more parameters (see Appendix 1). However, in molecular studies, authors rarely measure more than one parameter and therefore it is still very difficult to test the value of theoretical systems with respect to mutants affecting phyllotaxis. In fact, no known mutation modifies a single phyllotactic parameter like the angle of divergence, the plastochrone ratio, or parameter *b*. For example, the mutant *forever young* seems to affect the number of parastichies, the angle of divergence, the plastochrone ratio, and the angles between parastichies (Callos *et al.*, 1994). How can we measure these parameters on mutant plants? Certain parameters are easy to measure while others, such as the angle between two parastichies, are more difficult to determine. Thus, which parameters are involved in the generation of phyllotactic changes and their effects are still open questions. As Reinhardt and Kuhlemeier (2002) state: '*Do specific phyllotaxis genes exist?*' It appears that mutations affecting phyllotaxis are primarily involved in specifying the size of the apex or floral primordia (Jackson and Hake, 1999). On the other hand, there are mutants such as *terminal ear* and *periantha* that alter phyllotaxis without having other effects on meristem size and organisation (Veit *et al.*, 1998).

A change in the shape of the meristem can also affect phyllotactic patterns (Leyser and Fumer, 1992; Laufs *et al.*, 1998; Jackson and Hake, 1999; Itoh *et al.*, 2000; Nishimura *et al.*, 2002). Interestingly, the gene expression patterns specifically related to new organs on the apex first appear as a ring at the periphery of the SAM, not as spots (Kelly *et al.*, 1995; Green, 1998; Smith *et al.*, 2006b; Braybrook and Kuhlemeier, 2010; Traas, 2019). Therefore, the size of the primordium in relation to the form and the size of the meristem is an initial constraint for the propagation of the phyllotactic system.

Many mutations affecting meristem organisation can modify the value of phyllotactic parameters (e.g. divergence angle, plastochrone ratio). Molecular biology studies (Callos and Medford, 1994; Carpenter *et al.*, 1995; Coen *et al.*, 1995; Bradley *et al.*, 1996; Luo *et al.*, 1996; Cronk and Möller, 1997; Stieger *et al.*, 2002; Peaucelle *et al.*, 2011; Chitwood *et al.*, 2012; Kierzkowski *et al.*, 2013; Sassi and Vernoux, 2013; Guenot *et al.*, 2014) raise fundamental and important questions for understanding phyllotaxis. In their review, Reinhardt and Kuhlemeier (2002) clearly formulated the following questions regarding phyllotaxis: (1) How is the site of organ formation selected? (2) How are organs formed? (3) How do pre-existing organs influence the position of new organs? Phyllotactic models are generally developed and used to explain the propagation of established patterns, but, with few exceptions, do not explain how phyllotaxy arises *de novo*, which is the central problem of phyllotaxis (Green, 1987; Jackson and Hake, 1999). Can global models of phyllotaxis integrate the recent discoveries of molecular biology, and, on the other hand, is a molecular explanation of phyllotaxis possible without a global theoretical framework?

Phyllotactic Mutants

Comprehensive reviews such as those of Reinhardt and Kuhlemeier (2002) and Galvan-Ampudia *et al.* (2016) distinguished groups of mutants that have an effect on organ formation and phyllotaxis: mutants defective in meristem organisation; mutants defective in organ initiation; mutants defective in organ separation; and mutants with altered organ number or organ position. Mutations directly affecting the plant's phyllotaxis (spiral vs. whorled) seem relatively uncommon. Although it is possible to associate a mutation with a particular phyllotactic modification, without referring to a particular model of phyllotaxy it is very difficult to determine precisely which phyllotactic parameter is affected by the mutation. Molecular processes that regulate the timing and position of organ initiation need to be further elucidated. For example, in *clavata* 1 mutants of Arabidopsis, flowers have a higher number of organs in each of the four whorls, and can also present additional whorls. These changes are correlated in some cases with an increase in the size of the floral meristem (Clark *et al.*, 1993a,b, 1996). However, in Arabidopsis, a mutation in the gene *PERIANTHA* affects floral organ number without affecting organ type or the size of the floral meristem (Running and Meyerowitz, 1996). Can a specific mutation control the plastochrone ratio or the size of the meristem? To answer this question, it is necessary to analyse available data in a rigorous theoretical framework that will provide us with a better understanding of the quantitative rules involved in the unusual phyllotactic patterns characteristic of some mutants.

The regulation of the size of the shoot meristem in Arabidopsis depends on a feedback loop between the *CLAVATA* (*CVL*) and *WUSHEL* (*WUS*) genes (Schoof *et al.*, 2000; Sassi and Vernoux, 2013). Fujita *et al.* (2011) developed a mathematical model of the SAM based on the dynamics of *CLV* and *WUS* in the framework of an activator–inhibitor system, incorporating cell division and spatial constraints. Maize plants are characterised by a distichous phyllotaxis, where leaves are initiated individually, alternating from one side of the SAM to the other in a regular pattern. Greyson and Walden (1972) described the *ABPHYL* (abnormal phyllotaxis) syndrome in maize (*Zea mays*). The range of patterns in *abphyl* mutants extends from a typical phyllotaxis of distichous leaves to a decussate arrangement. In *abphyl* embryos, the SAM is much larger than in wild type embryos. The decussate shoot meristems are larger than normal throughout development, but the general structure and organisation of the meristem is not altered. According to Jackson and Hake (1999), the *ABPHYL1* gene regulates morphogenesis in the embryo, and participates in the phyllotaxis of the shoot. Braybrook and Kuhlemeier (2010) observed that *ABPHYL1* is central to the interplay between cytokinins, meristem size, and auxin/PIN1-determined phyllotactic patterning. In addition to the change in phyllotaxis, the leaves of *abph1* and *abph2* mutants are narrower than in normal plants (Yang *et al.*, 2015).

Jackson and Hake (1999) questioned whether phyllotaxis results from a change in the size of the meristem or if it affects the size of the meristem. In the *abphyl1* mutant, the meristem is enlarged in the embryo, before the first decussate pair of leaves is visible (Jackson and Hake, 1999), indicating that the size of the meristem is determined before the appearance of a phyllotactic pattern.

Because the SAM is larger in mutant plants, the plastochrone ratio is presumably smaller in these plants than in wild type plants. However, the quantitative relationship between the phyllotactic parameters in seedlings of the *abphyl* mutant plants, and the precise correlation between molecular data and phyllotactic changes, are not known. Itoh *et al.* (2012) identified *DECUSSATE (DEC)* as the gene responsible for phyllotactic pattern alterations in rice. Loss-of-function *dec* mutants showed a conversion from a typical distichous to a decussate phyllotactic pattern and had an enlarged SAM with enhanced cell division activity. Another gene affecting the size of the SAM (among other features) is *FASCIATED EAR4 (FEA4)* in maize (Pautler *et al.*, 2015), which is thought to promote differentiation at the periphery of the meristem by regulating auxin-based responses and genes associated with leaf differentiation and polarity. The *fea4* mutants have greatly enlarged vegetative and inflorescence meristems, which ultimately affect the patterning of organs.

Carpenter *et al.* (1995) performed a combination of morphological and genetic analyses to understand how the genes *FLORICAULA (FLO)* and *SQUAMOSA (SQUA)* interact to promote a change of phyllotaxis from spiral to whorled in the inflorescence of *Antirrhinum*. These authors proposed that a common process underlies the phyllotaxis of wild type, *flo*, and *squa* meristems development. However, the relative timing of primordium initiation and growth differed between these meristems, in terms of two independent events: (1) selecting the zones for potential primordium formation, and (2) partitioning these zones into discrete primordia. Failure of the second event to take place led to the formation of the continuous double spirals occasionally observed in *flo* mutants. The appearance of double helices in *Antirrhinum* is analogous to the biastrepsis of *Dipsacus sylvestris*, where a physical constraint on the stem causes the torsion of the stem (Barabé and Vieth, 1990; Vieth, 1998).

Liu *et al.* (1993) described the mutant embryos of Indian mustard *Brassica juncea* with fused cotyledons forming a tubular structure. The wild type of *Brassica* is characterised by spiral phyllotaxis and opposite cotyledons. These mutant embryos are phenocopies of the pin-formed (*pin 1-1*) mutants of Arabidopsis, with reduced auxin polar transport activity in the inflorescence axis, and of the *emb30 (gnom)* mutant embryos of the same species. The effect of fused cotyledons on the phyllotactic patterns of the seedlings remains to be described.

Otsuga *et al.* (2001) have shown that the gene *REVOLUTA (REV)* regulates meristem initiation in a lateral position in Arabidopsis. Within the inflorescence shoot meristem, *REV* expression appears to induce three to five incipient floral primordia on the flanks of the meristem. Thus, Otsuga *et al.* (2001) proposed that *REV* acts in a lateral position to activate the expression of known meristem regulators. Similarly, *Oryza sativa PINHEAD1* (*OsPNH1*) expression specifies the position of lateral organ primordia in the SAM of rice (Nishimura *et al.*, 2002). In the *leafy* mutants of *Arabidopsis*, floral organs are not whorled as in wild type flowers but intermediate between a spiral and a whorled pattern (Weigel *et al.*, 1992). These *leafy* mutants also validate the premise that the specifications of organ identity and organ position are two independent phenomena (Weigel *et al.*, 1992), suggesting that several genes are involved in the determination of phyllotactic patterns (Appendix 1).

Phyllotactic Mutations and Their Correlation with Phyllotactic Parameters

The *fey* mutation (*FEY* is *Forever Young*) results in a disruption of leaf positioning and meristem maintenance (Callos and Medford, 1994). Compared to the regular spiral phyllotactic pattern observed in wild type *Arabidopsis* where the divergence angle is approximately 137°, in the *fey* mutants, leaf primordia were initiated at abnormal positions with divergence angles varying from 29° to 176° (Callos and Medford, 1994). These authors also found that almost half (47%) of the *fey* leaf primordia were centric (i.e. lacked dorsiventral symmetry). However, it is not known if there is a link between this leaf form and signals originating from the SAM.

In the SAM of flowering plants, there is a close relationship between the angle of divergence and the leaf arc angle (i.e. the size of the foliar primordium) (Rutishauser, 1998). Although changes in the leaf arc could be associated with changes in organ positioning (divergence angle), whether or not this parameter plays a key role in determining phyllotactic patterns, or simply results from a pattern, has yet to be determined. As Callos *et al.* (1994) stated, the role of *FEY* in pattern formation could be direct (if it modified a factor involved in the phyllotactic organisation) or indirect (if it functioned to maintain SAM integrity).

Jean and Barabé (1998b) analysed the data presented in Table 4.1 of Callos *et al.* (1994) statistically, but found no significant correlation between the divergence angle and the plastochrone ratio, and no significant correlation between the leaf pair and the plastochrone ratio or the divergence angle. There were also no differences in terms of plastochrone ratio between the different mutants. However, representative *fey* apices #2 and #3 seemed to have an average divergence angle smaller than that of *fey* apices #1 and #4 and wild type. The mutant Arabidopsis plants show such great variation in the values of the phyllotactic parameters, that it is difficult to establish any type of quantitative correlation between them. Additionally, the few measurements available are not in agreement with the theoretical models that predict a strong correlation between the above parameters, although this might be related to small sample size.

Tamaoki *et al.* (1999) found that the overexpression of the gene *Neutral TreHalase* (*NTH1*) in the shoot apex of tobacco leads to an increase in the plastochrone ratio and to the appearance of an abnormally curved leaf. Moreover, the degree of the leaf curvature was positively correlated with the plastochrone ratio, suggesting that these parameters could be interdependent. Tamaoki *et al.* (1999) also suggested that the overall structure of the SAM is correlated with the curvature of the leaves in *NTH1*-transformed tobacco plants. This is not entirely surprising considering that there is a quantitative relationship between the size of the SAM and the leaf arc.

Although there was no significant difference in the divergence angle between wild type and transgenic plants, the plastochrone ratio was significantly greater in transgenic plants (Tamaoki *et al.*, 1999). This finding has two interpretations:

In theoretical models, the divergence angle and the plasto-chrone are quantitatively linked to the angle between opposed parastichies (γ) and the number of parastichies ($m + n$). Because the number of parastichies remains the same in mutants (Fig. 10 in Tamaoki *et al.*, 1999) Jean and Barabé (1998b) hypothesised that the angle between parastichies changes with the plastochrone ratio. However, further measurements are needed to confirm or disprove this hypothesis.

In theoretical models, the plastochrone ratio is directly pro-portional to the ratio between the size of the foliar primordia and the size of the meristem (parameter b). As discussed previously, parameter b corresponds to a biological constraint linked to the genotype of the plant. For transgenic tobacco, this is also sup-ported by the fact that the expression of leaf curvature requires an interaction between the leaf primordia and the SAM. There-fore, is there a difference in the value of parameter b between the non-transformants and the transformants?

An increase in the plastochrone ratio is associated with a decrease in the size of the apex. For example, in maize, the apex is smaller in distichous phyllotactic systems than in spiral phyllotac-tic systems or decussate-opposite phyllotaxis systems (Jackson and Hake, 1999). Thus, in plants overexpressing *NTH1*, it is expected to find a smaller SAM and a larger leaf base. These predictions, deduced from theoretical models, could be tested by searching for new mutations affecting phyllotactic parameters.

Itoh *et al.* (2000) described six *Shoot Organisation* (*sho*) mutants in rice, all exhibiting similar phenotypes: random phyllotaxis, short plastochrone ratio, aberrant leaves, and malformed SAMs. Because this gene regulates the SAM organisation and the pattern of leaf initiation, most of the leaves in the mutant plants develop into thread-like structures and do not encircle the SAM. This type of modification corresponds to a reduction in the leaf arc. In many cases, the SAM of the mutant plants was flat in relation to that of wild type plants.

Itoh *et al.* (2000) performed a correlation analysis between the variables of SAM geometry and the different phyllotactic parameters. The authors found no correlation between SAM volume and the phyllotactic variables (plastochrone, divergence angle, and entropy, defined as a measure of the degree of order). This suggested that SAM volume does not affect the leaf initiation pattern. However, SAM width was positively correlated with entropy and negatively correlated with the divergence angle. This is in agreement with the predictions of theoretical models, in which the angle of divergence decreases as entropy increases (Jean, 1994). Itoh *et al.* (2000) also suggested that SAM shape, rather than size, primarily affects the pattern of leaf initiation. Overall, there is no common theme or unifying principle for the manifestation of a phyllotactic pattern in mutants, and no correlation between standard phyllotactic parameters (i.e. angle of divergence, plastochrone ratio, number of parastichies; see the chapter on "Geometric Parameters").

Other Examples of Genes Affecting Phyllotaxis

Based on several studies, there appears to be an important correlation between the number of lateral organs and SAM size (Callos *et al.*, 1994; Jackson and Hake, 1999; Itoh *et al.*, 2000), and this was also observed in floral systems. For example, the enlarged *clv1* meristems of Arabidopsis were correlated with abnormal phyllotaxis and increased number of floral organs (Clark *et al.*, 1993a,b). A similar relationship was found for rice *floral organ number (fon)* mutants, where the number of floral organs increases as the SAM was enlarged (Nagasawa *et al.*, 1996). In contrast, *wus* and a weak allele of *shoot meristemless (stm)* led to a reduction in the size of the meristem and number of floral organs in Arabidopsis (Clark *et al.*, 1996; Laux *et al.*, 1996; Sassi and Vernoux, 2013). Other mutations, such as *periantha (pan)*, led to an increase in the number of floral organs without affecting the size and shape of the meristem (Running and Meyerowitz, 1996). However, the *PAN* gene, which regulates the number of organs by controlling the distance between two primordia (Chuang *et al.*, 1999), seems to affect the plastochrone ratio.

In the rice *pla1* mutant, the time span between the initiation of two successive leaf primordia (plastochrone) is modified, but not its phyllotaxy. According to Itoh *et al.* (1998), the rate of leaf production doubles in this mutant, and the SAM is enlarged without any change in shape, but the phyllotaxis remains distichous. Itoh *et al.* (2000) suggested the changes in SAM size only affected the rate of leaf production if there was no change in SAM shape.

In maize, Veit *et al.* (1998) have shown that the *terminal ear1 (te1)* mutants have an increased rate of leaf production, aberrant phyllotaxy, and abnormal phytomers. The *TE1* gene, therefore, appears to inhibit phytomer/leaf initiation in the SAM. Also, the SAM of the *te1* mutant had an abnormal geometry, producing leaf primordia '*higher on the apical dome than is normal*' (Veit *et al.*, 1998). Chaudhury *et al.* (1993) reported that the *altered meristem program (amp1)* mutants of Arabidopsis have altered plastochrone and phyllotaxis, but unfortunately the size and shape of the SAM were not described. Tamura *et al.* (1992) showed that the *organ differentiation mutant5 (odm5)* rice mutants displayed random phyllotaxy, short plastochrones, abnormal leaves, and malformed SAMs.

The overall structure of the SAM as a holistic entity is also crucial for maintaining the foundation on which phyllotactic patterns can develop. Genes like *HAIRY MERISTEM (HAM)* in *Petunia* species are essential for maintaining the integrity of the SAM (Stuurman *et al.*, 2002). When the structure of the SAM is altered or enlarged, as in the *clv* mutants of Arabidopsis, a disrupted phyllotactic pattern emerges, where the angle of divergence deviates from the wild type pattern of 138° within the first six rosette leaves (Leyser and Fumer, 1992; Clark *et al.*, 1993a,b; Bowman, 1994).

The *PLETHORA* gene of Arabidopsis, identified as a key regulator of *PIN1* activity, can also modulate shoot organ positioning on the SAM by affecting *PIN1* 'dosage' (Prasad *et al.*, 2011). Pinon *et al.* (2013) also determined that the activity of *PLETHORA* at the centre of the meristem is necessary for the manifestation of the typical spiral phyllotactic pattern. Other genes such as *BELL-RINGER* in Arabidopsis can also modulate phyllotaxis by affecting the expression of cell wall loosening enzymes at the site of incipient primordia (Peaucelle *et al.*, 2008, 2011). As suggested by Smith *et al.* (2006b), phyllotaxis is not 'governed by a single mechanism but represents a combined effect of several factors' operating at different hierarchical levels (i.e. from cells to organs).

Morphogens: The Role of Auxin

Over the last 50 years, a great number of experiments involving surgical alterations and/or the application of growth regulators (auxin and gibberellin) have been conducted for a better understanding of the establishment of phyllotactic patterns (Snow and Snow, 1931; Loiseau, 1969; Steeves and Sussex, 1989; Reinhardt *et al.*, 2000, 2005). For example, Marc and Hackett (1991) observed the conversion of the (2,3) pattern characteristic of mature apices to an incipient (1,1) juvenile pattern after gibberellin (GA_3) treatment in *Hedera helix*. Among the main hypotheses used to explain phyllotactic patterns, the field theories of diffusion (promoter or inhibitory), studied by many authors using Fick's laws (or similar laws in physics), highlight the idea that a morphogen is responsible for pattern formation (Jean, 1994). Mathematical modelling has shown how an asymmetrical pattern of distribution of a morphogen can generate different phyllotactic patterns (Meinhardt *et al.*, 1998; Smith *et al.*, 2006b). Several substances have been identified as potential morphogens in what is proving to be a relatively complex mechanism (Liu *et al.*, 1993; Reinhardt *et al.*, 1998; Stieger *et al.*, 2002; Smith *et al.*, 2006b).

In his model, Yotsumoto (1993) considered that the concentration of an inhibitor might fall to zero and therefore he was able to simulate unusual patterns, such as spiro-decussate, pseudo-decussate, and oscillation distichy, as well as transitions between patterns, but he could not reproduce superposed whorls, monostichy, and oscillation decussate. Using the *Nth1* mutant, Tamaoki *et al.* (1999) hypothesised that an inhibitory substance derived from leaf primordia would be a good candidate for explaining the direction of leaf curvature and the direction of phyllotaxy in tobacco plants. Meinhardt *et al.* (1998) were able to simulate different patterns of phyllotaxis by using a reaction–diffusion system. The postulated inhibitor or promoter morphogen was identified as the phytohormone auxin or a protein (expansin), both diffusing in the SAM. Currently, we know with great certainty that auxin regulates the initiation and radial position of leaves and flowers (Reinhardt *et al.*, 2000; Smith *et al.*, 2006b; Kuhlemeier, 2007). In their model of phyllotaxis derived from experimental data, Smith *et al.* (2006b) describe an auxin-transport patterning mechanism based on five basic facts regarding auxin (see detailed description below).

Reinhardt *et al.* (2000) initially proposed that auxin determines the radial position and the size of lateral organs, based on experimental work involving *Arabidopsis pin-formed1-1* mutants. The results obtained later by Stieger *et al.* (2002) suggested that functional auxin-concentration gradients determined the site of primordium formation. Both Reinhardt (2005a) and Smith *et al.* (2006b) outlined a mechanism of polar auxin transport in the SAM and the role of *PIN* proteins in generating 'peaks' of auxin concentration that ultimately determine the site of organ formation (Figs. 4.2 and 4.3)

Building on the discovery that leaf positioning around the stem is regulated by auxin transport (Reinhardt *et al.*, 2000), the theoretical model proposed by Smith *et al.* (2006b) using computer simulations showed that the interaction between auxin and its transporter (PIN) regulates the production of different regular

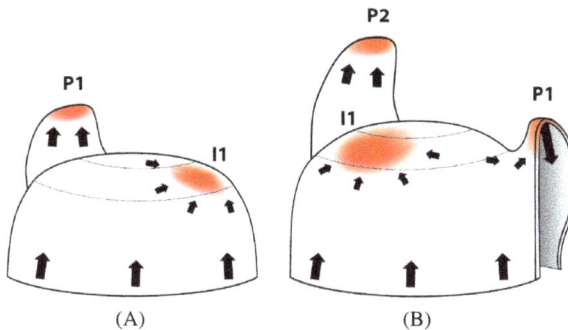

Fig. 4.2. Conceptual model of the regulation of phyllotaxis by polar auxin fluxes in the shoot apical meristem. Adapted from Reinhardt *et al.* (2003). (A) PIN1 orientation directs auxin fluxes (arrows) in the L1 layer, leading to the accumulation of auxin (red) at the initiation site (I1) in the peripheral zone. This accumulation eventually results in organ induction. (B) Later, basipetal PIN1 polarisation inside the bulging primordium (P1) drains auxin into inner layers, depleting the neighbouring L1 cells. As a consequence, another auxin maximum is created in the peripheral zone at position l1 removed from primordia P1 and P2. (From Reuihardt *et al.*, 2003; reprinted with permission from Springer Nature [as modified in Smith *et al.*, 2006].)

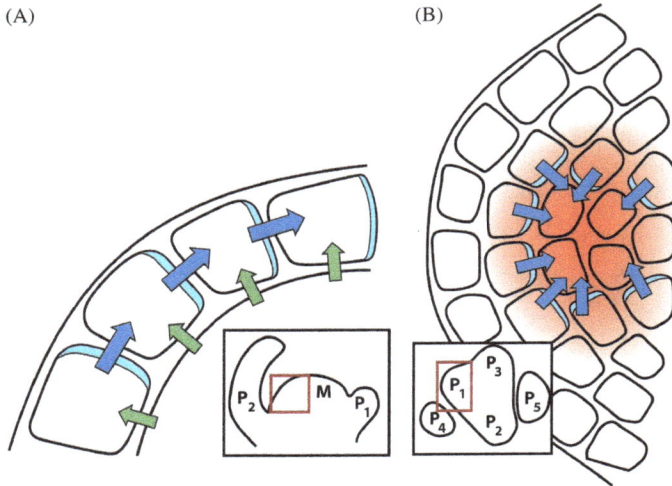

Fig. 4.3. **(A)** Auxin transport in the epidermal layer L1 of the shoot apical meristem. L1 cells express the presumptive auxin-influx carrier AUX1, leading to the accumulation of auxin in L1 (green arrows). PIN1 is also expressed in the L1 layer, where it is located to the upper side of the cells (light blue). This results in the acropetal transport of auxin towards the meristem centre (blue arrows). Inset: The depicted area in the context of the apex. **(B)** Auxin accumulation in young primordia. PIN1 is induced in young primordia. It becomes localised to the side of the cells that points to the centre of the primordium (light blue). This results in the accumulation of auxin in the primordium and its withdrawal from the surrounding cells (blue arrow). The resulting auxin gradient (red) confers positional information to the cells, thus allowing them to establish open and boundary identity. Inset: Location of the P1 position in the context of the apex. (From Reinhardt, 2005b; reprinted with permission from Elsevier.)

phyllotactic patterns. Later, Bayer *et al.* (2009) were able to develop an integrated model that provides a three-dimensional perspective of the developmental patterns for phyllotaxis and midvein formation. The role of the developing midvein and its effect on the stability and overall order of the phyllotactic pattern was recently further elucidated by Deb *et al.* (2015). Using laser ablation, these authors showed that the removal of an incipient midvein led to a transient change in phyllotaxis caused by a transient accumulation of auxin in the overlying primordium.

Experimental data showed that an increase in the concentration of auxin in the L1 layer of the SAM appeared concomitantly with the presence of the auxin-transport protein PIN1 in the direction of the centre of the incipient primordium (Bayer *et al.*, 2009). This auxin-efflux carrier is directionally localised to the plasma membrane of cells with higher auxin concentration. This process, called 'up-the-gradient' polarisation, increases the local concentration of auxin resulting in a distinct auxin peak where the leaf primordium will be initiated. Cycles of increases and decreases in auxin concentration are thought to direct primordia through the different stages of development (Heisler *et al.*, 2005).

The initiation of the first vascular bundle occurs approximately at the same time and in the same zone as the initiation of the leaf primordium (Bayer *et al.*, 2009). The earliest molecular markers for the formation of the first vascular bundle, *PIN1* and auxin reporter *DR5*, are the same as those involved in phyllotactic patterning (Scarpella *et al.*, 2006). This suggests that both phyllotaxis and midvein formation are based on directional auxin transport (Bayer *et al.*, 2009) and drainage through the leaf primordia (Deb *et al.*, 2015).

The concept of canalisation was proposed many years ago to explain the formation of veins in plants (Mitchison, 1980, 1981; Sachs, 1981). According to this concept, the ability of cells to transport auxin increases with auxin flux. When auxin flows through a plant tissue it follows a given path that will be reinforced by a flux-based feedback process, which leads to the formation of well-defined high-flux pathways that are the precursors of veins. In molecular terms, this concept implies that the direction of the auxin-flux controls the polarised localisation of the auxin carrier PIN1, which in turn enhances the flux in that direction, a process that Bayer *et al.* (2009) named 'with-the-flux polarization'. Bayer *et al.* (2009) were able to integrate the patterning mechanisms proposed for phyllotaxis (up-the-gradient) and the processes involved in the formation of vascular bundles (with-the-flux) in the same model and showed that these mechanisms operate concurrently.

This model is based on experimental data, some of which are highlighted in Fig. 4.4, and described eloquently by Bayer *et al.* (2009, p. 176) as follows: '... the process of leaf position begins with the convergent localisation of PIN1 proteins and an auxin maximum at I1. While auxin in this convergent point continues to accumulate, the central L1 cells begin to display basal PIN1 polarisation. PIN1 turns on in the internal layers in a triangular expression domain that gradually resolves into a narrow canal with high auxin concentration and basal PIN1 polarisation in the centre and lateral polarisation in the adjacent cells. Thus, it appears that cells

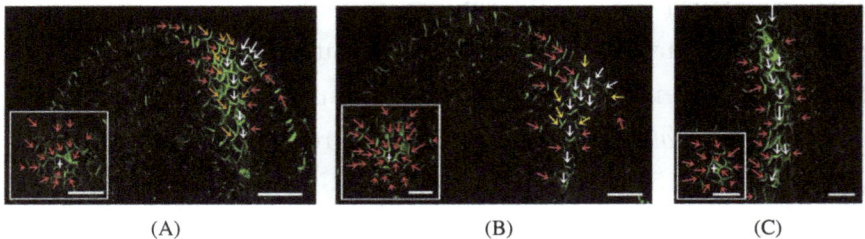

(A) (B) (C)

Fig. 4.4. Progressive establishment of basal and lateral PIN1 polarisation during midvein formation. PIN1 immunolocalisation on vegetative tomato meristem sections, visualised by confocal imaging. Longitudinal and transversal sections (insets) of an incipient primordium (A), bulging leaf primordium at early P1 stage (B), and primordium at P2 stage (C). PIN1 polarisation is indicated by arrows (red for lateral towards the future midvein, yellow for oblique, and white for basal). The white star indicates the midvein. Bar = 20 μm. (From Bayer *et al.*, 2009.)

can change their polarisation from convergent (anticlinal) to basal in the L1, and adjacent cells in inner tissue that were previously uncommitted can adopt lateral and basal polarities'.

This model is based on the following assumptions: (1) *PIN1* is induced by auxin; (2) midvein formation is initiated by an increase in auxin concentration; (3) the cells in L1 have a different identity from cells in the inner tissues; (4) the up-the-gradient and with-the-flux polarisation mechanisms act simultaneously in varying proportions within individual cells; and (5) the transition from up-the-gradient to with-the-flux polarisation depends on auxin concentration. In this model, the feedback interaction between auxin concentration, flux, and PIN1 localisation directs indole-3-acetic acid (IAA) to the site of primordium initiation in L1, and then into the sub-epidermal tissue layers. This basal polarisation represents the beginning of midvein initiation which ultimately connects with the stem.

Expansins and Auxins: A Bridge Between Different Models?

Expansins are extracellular proteins that increase plant cell wall extensibility *in vitro* and are thought to be involved in cell expansion. Local induction of ectopic expansin expression in the meristem induces the initiation of leaf primordia (Pien *et al.*, 2001). It has been shown that the exogenous application of purified expansin protein can induce local growth of the meristem, resulting in the formation of bulges similar to early primordia (Fleming *et al.*, 1997). Expansins are also believed to play a role in the control of organ size and morphology (Cho and Cosgrove, 2000), and they seem to primarily affect morphogenesis via changes in biophysical stress patterns in the meristem (Fleming *et al.*, 1999). There is a strong correlation between expansin gene expression and growth rate, but this correlation is not absolute (Caderas *et al.*, 2000).

A localised increase in expansin gene activity induces a bulging of the tissue at the site of leaf initiation (Fleming *et al.*, 1999), which is the primary event of foliar morphogenesis. In the model of Fleming *et al.* (1999), changes in expansin concentrations would lead to changes in physical stress necessary to induce the buckling of surface layers of the shoot apex. In turn, the increased concentration of expansin in the internal corpus would allow the cells of the corpus to respond to the variations in stress parameters by expanding and dividing to occupy the space left by the tunica as it bulges upwards.

Reinhardt *et al.* (1998), by examining the role of expansin in leaf initiation, attempted to establish a link between molecular mechanisms and the hypotheses of Snow and Snow (first available space), Green (buckling of a growing sheet), and Meinhardt (reaction–diffusion of morphogens; existence of inhibitory substances) regarding phyllotactic organisation. Expression of the expansin gene *LeExp18* in tomato (*Solanum lycopersicum*) leads to bulging, which in turn activates a new morphogenetic pathway (Reinhardt *et al.*, 1998). According to these authors, signals (chemical gradient, physical forces or both) originating from existing primordia determine the position of new primordia. The expression of *LeExp18* at the site of initiation increases tissue extensibility and will induce bulging. It is interesting to note that the local increase in *LeExp18* expression does not coincide with increased cell division activity.

The pattern of *LeExp18* expression (induced by IAA) is a manifestation of the regulation of phyllotaxis in the meristem. It constitutes a positive molecular marker for leaf initiation (Reinhardt *et al.*, 1998) in addition to the activities of the auxin efflux and influx carriers, namely PIN and AUX/LAX proteins (Stieger *et al.*, 2002; Kramer, 2004). The promotion of the extensibility of the cell wall induces a deformation of the surface of the meristem leading to the formation of a bulge (Reinhardt *et al.*, 1998). In this situation, there is an analogy with the model of Green (1992a), which postulates that a new bulge will appear where the surface tension is at a minimum. Nakayama *et al.* (2012) also showed that mechanical strain could affect auxin movement and accumulation in the SAM of tomatoes, more specifically of PIN1 efflux carriers. This information, in combination with the model proposed by Reinhardt *et al.* (2003), whereby interactions between existing and incipient primordia in the SAM are mediated by auxin, outlines the numerous factors that come into play at the cellular and organ levels. The increased use of loss-of-function and knock-out mutants will likely provide further insight into the signalling pathways involved in the generation and maintenance of phyllotactic patterns.

A recent paper by Bhatia *et al.* (2016) revealed that *MONOPTEROS* (*MP*) is involved in the formation of primordia and ultimately in the manifestation of phyllotactic patterns. The authors proposed a mechanism in which auxin-regulated MP orients the polarity of PIN1 in adjacent cells to promote the efflux and further local accumulation of auxin, thereby forming a positive feedback loop that leads to organ formation. This is in accordance with other models of phyllotaxis where the polarisation of PIN1 in neighbouring cells responds to high intracellular levels of auxin. The results by Bhatia *et al.* (2016) also supported the existence of a mechanical stress feedback model where auxin triggers the expression of cell wall loosening enzymes, thereby increasing the tensile stress of adjacent cell walls and the polarisation of PIN1 protein to these areas. The authors suggested that MP, therefore, plays 'an instructive rather than permissive role' in specifying the position of a lateral organ, and showed that MP activity in sub-epidermal layers is essential to outline and 'stabilise' the distribution patterns

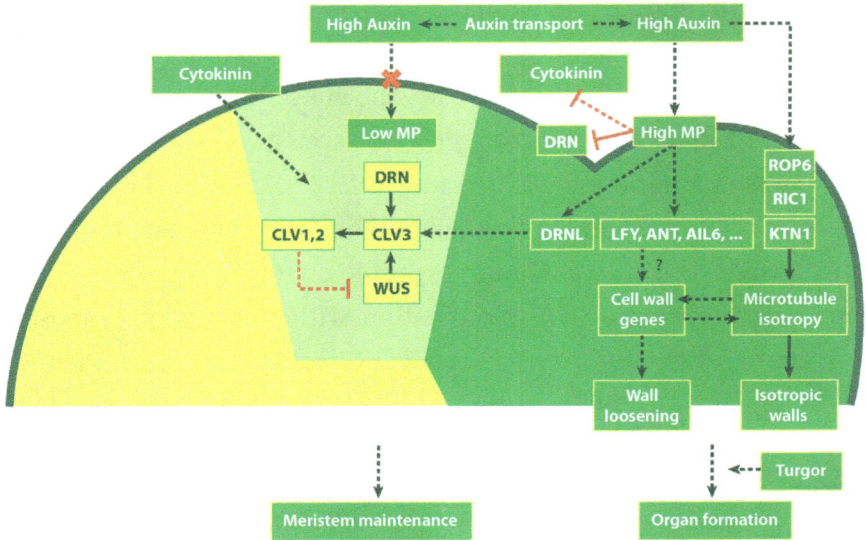

Fig. 4.5. Molecular and cellular regulation of organ initiation at the periphery of the meristem. Auxin transport leads to maximum auxin concentration at the centre (light green area) and periphery (dark green area) of the meristem, but because the centre is relatively insensitive to auxin (red cross), its effects seem to be limited and cytokinin-driven meristem maintenance dominates. The increased auxin concentration at the periphery leads to cell wall loosening and cell isotropy, which involves both transcriptional and cellular responses. Depending on their wall properties, cells will grow at particular rates and in particular directions driven by turgor pressure. Dotted arrows represent indirect effects and solid lines direct relationships. Green arrows stand for positive control and red lines for inhibitions. (From Trass, 2019.)

of auxin. A recent review of the molecular and cellular regulation of organogenesis at the SAM by Traas (2019) supports this theory (Fig. 4.5).

During the last decade, several models were developed based on the active polarised transport of auxin among cells (Heisler and Jönsson, 2006; Jönsson *et al.*, 2006; Smith *et al.*, 2006b; Stoma *et al.*, 2008). Compared to the reaction–diffusion models, these polar transport models are based on cell-to-cell signalling allowing the coordination of PIN1 (efflux carrier of auxin) polarity towards neighbouring cells containing the highest concentration of auxin. Two general types of polar auxin-transport models were applied to study phyllotaxis: concentration-based models and flux-based models. In concentration-based models, PIN levels increase on the cellular membrane facing the neighbouring cell with the highest concentration of auxin, up to the gradient. Flux-based models show how small fluctuations in the concentration of auxin may be amplified in the distinct streams where PIN1 is transported in the direction of the flux. In these two types of models, the amplification of a small local increase in auxin concentration induces the production of a new primordium, while simultaneously depleting auxin in neighbouring cells, resulting in the initiation of a new primordium at a fixed distance from the preceding one. Recently, the intracellular partitioning model was proposed to explain PIN1 polarity (Abley *et al.*, 2013). This third model does not require the establishment of differential auxin concentrations nor auxin flux to obtain cell polarity (Bhatia and Heisler, 2018), but it has not been used in the context of the generation of phyllotactic patterns. However, Bhatia and Heisler (2018) performed a detailed comparison of the three types of models.

The following sections of this chapter outline the three molecular models of Smith *et al.* (2006b), Jönsson *et al.* (2006), and Stoma *et al.* (2008).

Comprehensive Auxin Distribution Model: The Work of Smith *et al.* (2006b)

Smith *et al.* (2006b) developed a computer simulation model based on molecular data that reproduces spiral, distichous, decussate, and tricussate patterns. They were also able to reproduce the frequently observed transition from decussate to spiral phyllotaxis.

This model, which is based on molecular mechanisms and computational algorithms, links the morphogenetic processes involved in plant development with the resulting geometry of phyllotactic patterns. According to this model, it is the interaction between two molecules, which depends on active transport between adjacent cells rather than a purely diffusion-based process, as it is the case in reaction–diffusion models (e.g. Meinhardt *et al.*, 1998), that explains the generation of the phyllotactic pattern.

As mentioned above the model is based on a series of assumptions stemming from experimental results underlying biological processes (Table 4.1). These assumptions correspond to the three components of the model: geometry of the meristem (i) and (ii); cell polarisation and auxin transport (iii)–(vi), and the emergence of phyllotactic patterns (vii)–(x). They are based on well-established facts related to the movement and function of auxin within the context of organ initiation at the SAM.

Table 4.1.   Assumptions of the comprehensive auxin distribution model for phyllotaxis

Pattern formation is a dynamic process occurring at the shoot apical meristem (SAM).
L1 is a channel for auxin transport in the SAM.
Auxin is locally produced or flows from lower parts of the shoot into L1.
Diffusion and active transport redistribute auxin in L1.
Auxin up-regulates PIN1 concentration in a cell.
The concentration of auxin in neighbouring cells determines PIN1 localisation.
Lateral organ initiation in response to auxin occurs only in the peripheral zone.
A new primordium is initiated where the concentration of auxin is highest.
New primordia may modify model parameters as they acquire a new developmental identity.
Primordia are sinks for auxin in L1 and auxin is translocated to internal layers to initiate vascular tissues.

Adapted from Smith *et al.* (2006b).

Equations governing cell polarisation and auxin transport

In this section, we expand on the mathematical basis of the model of Smith *et al.* (2006b). In this model, cell polarisation and auxin transport are represented as an interaction between four factors: (1) overall concentration of PIN1 within the cell, (2) allocation of PIN1 proteins to individual cell membranes, (3) transport of IAA (auxin) between a cell and its neighbours, and (4) changes in IAA concentration. This system is based on the fact that the concentration and localisation of PIN1 proteins depend on IAA concentrations in a cell and its neighbours. Consequently, the two primary variables involved in this model are the concentration of PIN1 proteins and IAA in an individual cell i.

Based on previous biological assumptions, Smith *et al.* (2006b) deduced that the concentration of PIN proteins in cell i can be calculated as

$$\frac{d[\text{PIN}]_i}{dt} = \text{Production} - \text{Decay}$$

$$= \frac{\rho_{\text{PIN}_0} + \rho_{\text{PIN}}[\text{IAA}]_i}{1 + \kappa_{\text{PIN}}[\text{PIN}]_i} - \mu_{\text{PIN}}[\text{PIN}]_i, \qquad (1)$$

where ρ_{PIN_0} is the initial production of PIN, ρ_{PIN} is a coefficient representing the upregulation of PIN1 production by auxin, κ_{PIN} controls the saturation of PIN1 production at high concentration, and μ_{PIN} is the decay constant. The presence of $[\text{IAA}]_i$ in the first term of the equation reflects the interrelation between the concentrations of PIN1 and IAA in the cell.

The changes in IAA concentration in cell i are given as

$$\frac{d[\text{IAA}]_i}{dt} = \text{Production} - \text{Decay} + \text{Diffusion}$$

$$+ \sum_j (-\text{Active\_transport}_{i \to j} + \text{Active\_transport}_{j \to i}), \qquad (2)$$

where the first three terms (Production – Decay + Diffusion)

$$= \frac{\rho_{IAA}}{1 + \kappa_{IAA}[IAA]_i} - \mu_{IAA}[IAA]_i - \sum_j D_j([IAA]_i - [IAA]_i), \quad (3)$$

and

$$\text{Active\_transport}_{i \to j} = T[PIN]_{i \to j} \frac{([IAA]_i)^2}{1 + \kappa_T([IAA]_j)^2}. \quad (4)$$

In Eq. (3), as the auxin concentration increases, the rate of auxin production decreases from the maximum value ρ_{IAA} and the decay constant κ_{IAA} regulates saturation and auxin decay. Diffusion occurs directly between neighbouring cells (intercellular space is ignored), where cell membranes act as the main limiting factor to diffusion. Smith *et al.* (2006b) considered that each diffusion coefficient D_j is proportional to the length of the membrane between a cell and its neighbouring cell j.

Equation (4) is related to active_transport$_{i \to j}$ and models the effect of PIN1 proteins in auxin movement from cells i to j. Here, T is a polar transport coefficient and κ_T is a transport saturation coefficient. In Eq. (1), $[PIN]_{i \to j}$ is the number of PIN1 proteins located in the membrane of cell i near the wall that separates cells i and j, and it is calculated as

$$[PIN]_{i \to j} = \frac{[PIN]_i \, l_{i \to j} b^{[IAA]_j}}{\Sigma_j l_{i \to j} b^{[IAA]_j}}, \quad (5)$$

where $l_{i \to j} = l_{j \to i}$ is the length of the wall that separates cells i and j. b controls the extent to which the PIN1 protein is affected by neighbouring cells.

Emergence of phyllotactic patterns

To generate phyllotactic patterns, Smith *et al.* (2006b) extended the active transport-based model by adding two factors that play a role in the phyllotactic organisation: one related to the geometry of the apex and the other, to IAA concentration. These factors are related to key assumptions vii–x (Table 4.1).

The surface of the apex is divided into three zones: central, peripheral, and proximal (below the peripheral zone). No auxin production occurs in the central zone. However, the coefficient controlling auxin production (ρ_{IAA}) is greater in the peripheral zone than in the proximal zone.

In the peripheral zone, where the primordia are produced, a primordium is initiated when the IAA concentration in two adjacent cells reaches a predefined threshold *Th*. After initiation, the radius *r* of the primordium increases with time following a predefined function.

In the model of Smith *et al.* (2006b), the development of a primordium has three consequences:

Higher auxin production occurs in differentiated cells, which is modelled by the formula

$$\text{Additional production} = \frac{\rho_{IAA}(\text{primordium})}{1 + \kappa_{IAA}[\text{IAA}]_i}\left(1 - \frac{d_i}{r}\right) \qquad (6)$$

where d_i is the distance from the centroid of primordium cell *i* to the centre of the primordium.

The increased polarisation of PIN1 proteins towards the primordium, compared to the zone without primordia. To represent this polarising factor, Smith *et al.* (2006b) used an auxin-concentration equivalent, [IAA'], instead of the current auxin concentration [IAA]. The auxin-concentration equivalent is given by

$$[\text{IAA}']_i = \max\left\{[\text{IAA}]_i, [\text{IAA}]_{\max}\left(1 - \frac{d_i}{r}\right)\right\}, \qquad (7)$$

where $[\text{IAA}]_{\max}$ is the model's maximum IAA concentration.

It is also assumed that excess auxin flows into the tissues of the primordium to initiate the vasculature system, which was demonstrated experimentally in tomato (Deb *et al.*, 2015).

By using this model, Smith *et al.* (2006b) generated distichous, decussate, tricussate, and spiral phyllotactic patterns. They noted that multiple parameter values need to change to simulate different phyllotactic patterns. For example, the change from distichous to decussate involves increasing IAA production, decreasing the width of the peripheral zone, and increasing the size of the central zone. They also noted that the involvement of many parameters in phyllotactic changes may explain why it is so difficult to produce mutants representing the transformation of a stable phyllotactic pattern into another.

Auxin Transport Between Cell Compartments: Model of Jönsson *et al.* (2006)

The model of Jönsson *et al.* (2006) is also based on recent empirical data from previous molecular biology studies in which the role of the localised auxin concentration in plant organ positioning and the mediator of auxin transport, the PIN1 protein, have been confirmed empirically. The authors used confocal microscopy images to quantitatively compare their theoretical model to experimental data and to estimate parameters involved in their model.

This model consists of the following: passive and active auxin transport; PIN1 as the mediator of auxin flux; auxin regulating the polarisation of PIN1, which cycles between internal and membrane compartments; and changes in the relationship between adjoining cells as a result of growth and mechanical constraints. The main mechanism features auxin and is similar in nature to the model of Smith *et al.* (2006b). A crucial part of this system is a feedback mechanism whereby cells can increase their auxin content by influencing PIN1 polarity in neighbouring cells and directing the net flow of auxin away from them. This was demonstrated clearly in simulations on a theoretical meristem using a cell-based model (Jönsson *et al.*, 2006).

Auxin transport model

Jönsson *et al.* (2006) developed a detailed model of cell–cell auxin transport. In this model, auxin can passively or actively cross a cellular membrane. Under this system, the net auxin flux is given by $J_{i \to j}^{\text{tot}} = J_{i \to j} - J_{j \to i}$, where each term includes passive and active transport. From this principle, they developed a general equation on which simulations are based, representing the net auxin flux, $J_{i \to j}^{\text{tot}}$, between two compartments (i, j) separated by a membrane:

$$J_{i \to j}^{\text{tot}} = D\left(A_i - A_{ij}\right) + T\left(P_{ij} \frac{A_i}{K_A + A_i} - P_{ji} \frac{A_{ij}}{K_A + A_{ij}}\right), \qquad (8)$$

where $A_i(A_j)$ is the auxin concentration in compartment $i(j)$, P_{ij} and P_{ji} are the PIN1 concentrations on opposite sides of the membrane between the two compartments (Fig. 4.6), D is the strength of passive transport, and T is the strength of PIN1-dependent active transport. The authors considered that active transport can

(A)

(B)

(C)

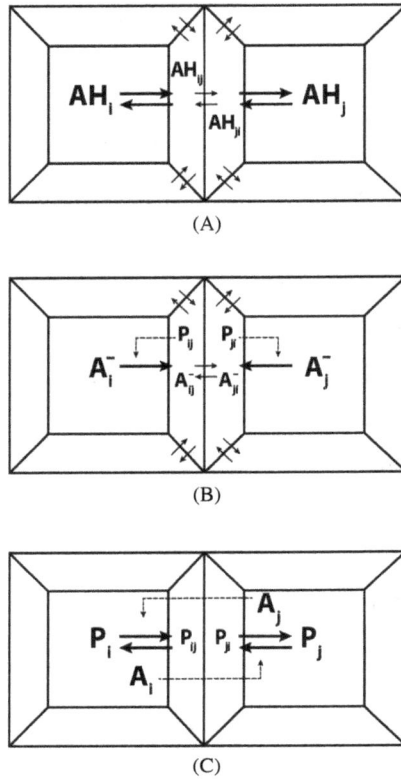

Fig. 4.6. Illustration of auxin transport and PIN cycling models. (A) *AH* (weak acid form) transport. (B) A^- (anion form) transport. Note that A^- influx is dependent on P_{ij}. This rate is low compared with the efflux, and this mechanism is not illustrated. (C) PIN1 cycling model. (From Jönsson *et al.*, 2006, supporting Fig. 7; copyright [2006] National Academy of Sciences.)

become saturated and can be modelled in a Michaelis-Menten formalism, where K_A is the Michaelis-Menten constant.

To test the prediction that cells forming new primordia have a relatively high auxin concentration, Jönsson *et al.* (2006) compared the theoretical model to experimental data of PIN1 localisation using quantitative confocal microscopy. To test the concordance between theoretical predictions and experimental data, they used a more detailed formulation of Eq. (8) in which parameter values are estimated experimentally. In this detailed model, the cellular compartments are surrounded by wall compartments between each neighbour and separated by a membrane (Fig. 4.6). Based on this context, the equation modelling the net auxin flux, $J_{i \to ij}^{\text{tot}}$, between a cellular compartment, i, and its neighbouring wall compartment, ij, corresponds to

$$J_{i \to ij}^{\text{tot}} = \left(D_{\text{efflux}} A_i - D_{\text{influx}} A_{ij} \right) + P_{ij} \left(T_{\text{efflux}} \frac{A_i}{K_A + A_i} - T_{\text{influx}} \frac{A_{ij}}{K_A + A_{ij}} \right), \quad (9)$$

where P_{ij} is the PIN1 located at the membrane.

Jönsson *et al.* (2006) considered that auxin is present in two forms within the plant: weak acid (AH) and anion (A^-). The cellular efflux consists of a passive transport rate corresponding to $D_{\text{efflux}} = p_{AH} f_{AH}^{\text{cell}}$ and an active transport rate corresponding to $T_{\text{efflux}} = p_A - f_{A^-}^{\text{cell}} N_{\text{efflux}}$. f_{AH}^{cell} and $f_{A^-}^{\text{cell}}$ are the fractions of the different auxin variants in the cell, p_{AH} and p_{A^-} are the membrane permeabilities, and N_{efflux} is the factor for efflux across the charge membrane. The influx from the wall to the cell can be passive and active as respectively expressed by $D_{\text{influx}} = p_{AH} f_{AH}^{\text{wall}}$ and $T_{\text{influx}} = p_A - f_{A^-}^{\text{wall}} N_{\text{influx}}$, where the individual parameters are defined above. By using this detailed model, the authors simulated the mechanism by which a new visible primordium has a clear auxin concentration peak.

Auxin-driven PIN1 cycling

The central hypothesis of Jönsson *et al.* (2006) emphasises the fact that the difference between the relative auxin concentration in adjoining cells determines where PIN1 will be active in the appropriate portion of the membrane between neighbouring cells. To analyse this auxin transport process, they simplified the cell-based model to a single variable by assuming that PIN1-mediated transport is unsaturated. This leads to the following equation:

$$\frac{dA_i}{dt} = D \sum_{k \in' N_i} (A_k - A_i) + T \sum_{k \in' N_i} (A_k P_{ki} - A_i P_{ik}) \tag{10}$$

In this equation, $A_i(A_k)$ is the auxin concentration in compartment $i(k)$, P_{ik} and P_{ki} are respectively the PIN1 concentrations on the membrane in compartments i and k, D is the strength of passive transport, and T is the strength of PIN-dependent active transport. The summation is over the set 'N_i. Jönsson *et al.* (2006) further simplified[1] Eq. (10),

---

[1] Jönsson *et al.* (2006, p. 1635) stated 'We assumed that the total amount of PIN1 in the cell and its membrane is constant and equal for all cells ($P = P_i^{tot} + \Sigma_k P_{ik}, \forall i$). We used a linear polarization feedback [$f(A_j) = k_1 A_j$] in Eq. (11) (see below, note 2) and assume that the PIN1 is in its equilibrium polarization state at any given time, which leads to $P_{ij} = PA_j/[(k_2/k_1) + \Sigma_k A_k]$. Finally, we assumed that most PIN1 is situated at the membrane ($k_2 \ll k_1$), which leads to $P_{ij} = A_j/(\Sigma_k A_k) \ldots$'

Jönsson *et al.* (2006) performed PIN1-cycling optimisation by simulating the detailed auxin model (Eq. (11)) with experimental values of PIN1 concentrations. For this purpose, they used equilibrium auxin values for optimising the PIN1 cycling model parameters so that they agreed with the empirical values. The predictions based on this model seem to agree with experimental data for PIN1 localisation obtained by confocal microscopy, indicating that cells forming new primordia have relatively high auxin concentration.

The hypothesis for PIN1 cycling is that auxin in a neighbouring cell (A_j) induces cycling from the cellular compartment (P_i) into the membrane located toward the neighbouring cell (P_{ij}). This process with a constant internalisation is described by the equations

$$\frac{dP_{ij}}{dt} = f(A_j)P_i - k_2 P_{ij}$$

$$\frac{dP_i}{dt} = \sum_{k \in' Ni} (k_2 P_{ik} - f(A_K)P_i), \tag{11}$$

where $f(A_i)$ encodes the feedback from auxin in the neighbouring cell and should be an increasing function of A_j, and k_2 is a constant. The summation is over the set of cell neighbours 'N_i for cellular compartment i. The authors use a which led to a model that depends only on auxin concentration and the parameters D, T, and P, where P is the constant total amount of PIN1 in the membrane and the cell ($P = P_i^{tot} + \Sigma_k P_{ik}, \forall i$). These three parameters can be related by the unique parameter D/TP.

This simple model, where auxin regulates its own polarised transport by a feedback phenomenon, can lead to the appearance of auxin peaks regularly distributed on a theoretical SAM by varying the D/TP value. For example, by keeping TP fixed, an increase in D (passive transport) leads to larger distances between peaks, and large enough values of D/TP lead to no emergence of a pattern from the model. This is all based on the fact that a cell with higher auxin concentration than its neighbour induces an increase in PIN1 localisation at the membrane of the neighbour towards itself, resulting in more auxin transport to itself. If this feedback system is strong enough, spatial patterns of auxin concentration emerge. Consequently, as a local maximum in concentration is established, the cells in the area deplete auxin in surrounding cells. However, at a certain specified distance from the maximum to a new maximum, cells will tend to have less auxin than their neighbours on either side. These cells will then start to transport auxin away from the initial maximum to the new maximum, resulting in a spatial pattern of peaks and valleys of auxin concentration.

Phyllotactic patterns

To simulate phyllotactic patterns (Fig. 4.7), Jönsson *et al.* (2006) used the cell-based auxin model Eq. (4) in a system allowing for

---

linear description, $f(A_j) = k_i A_j$, as well as a saturable form allowing for nonlinear feedback and described by a Hill-type equation, $f(A_j) = k_1 \left[\frac{A_j^n}{K^n + A_j^n} \right]$, where k_1 is the maximal rate and K and n are the Hill constant and coefficient, respectively (Jönsson *et al.*, 2006, p. 1638).

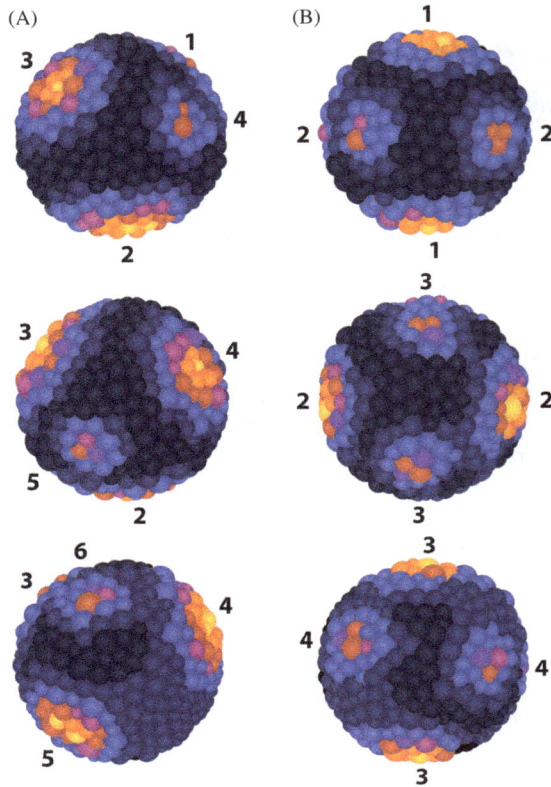

Fig. 4.7. Simulation of the phyllotaxis model on a half-sphere cylinder surface including cellular growth and proliferation. The main image shows a top view (insets show a side view) and time increases from top to bottom. The two simulations have different values for the size of the defined central zone. (A) Peaks formed in a spiral pattern. (B) Peaks formed in a whorled (decussate) pattern. (Modified from Jönsson *et al.*, 2006; copyright [2006] National Academy of Sciences.)

the growth and division of theoretical cells on a half-sphere surface connected to a cylinder. PIN1 polarisation is based on a linear relationship describing auxin dependence $f(A_j) = k_1 A_j$. They assumed equilibrium concentrations leading to

$$P_{ij}^* = \frac{k_1 A_j P_i^{\text{tot}}}{K_2 + \dfrac{k_1}{|N_i|} \sum_{k \in N_i} A_k}. \tag{12}$$

where P_i^{tot} is the total amount of PIN1 in the cell. They also assumed that it is constant in their simulations. $|N_i|$ is the number of neighbours that enter as a spatial contribution to the surrounded cell.

In the model, when a new auxin peak is produced in cells that become further removed from the SAM, the auxin concentration in the apex decreases quickly unless new auxin is supplied.

To avoid the formation of peaks in the apical zone, Jönsson *et al.* (2006) defined a central and a peripheral zone. For this purpose, they hypothesised that a molecule X is produced only outside the apical zone and induces auxin production. They also postulated that cells are spherical and have a uniform radial growth rate at the level of the apex and that the mechanical interactions between them are modelled based on repulsive spring forces between neighbouring cells. According to this model, cells of the apical region divide when they reach a certain threshold size, and they are eventually displaced outside the apical region towards the periphery. By using this system, the authors simulated spiral-like patterns in which a new auxin peak is formed at each sequential point in time and decussate-like patterns in which the peaks form in pairs in alternate positions. In this simulation, the size of the defined central zone is responsible for the difference between the two patterns.

Merks *et al.* (2007) proposed an alternative mechanism for polar auxin transport derived from the model of Jönsson *et al.* (2006). They used a travelling-wave system to explain the canalisation of auxin; however, they did not study it in the context of phyllotaxis.

Heisler and Jönsson (2006) further elaborated a numerical model of auxin transport based on a feedback mechanism and incorporated the auxin influx carrier AUX1, which promotes intracellular auxin accumulation. However, they performed simulations of their model on a two-dimensional surface of cells instead of a three-dimensional one, as was the case in Jönsson *et al.* (2006).

Heisler and Jönsonn (2006) also stated that the parameter values are mostly based on experimental data.[2] However, they considered nearly 30 parameters in their equation; we believe that the more we want to take into account biological processes the more we need to include parameters for which it is not always easy to find an empirical value.

Further, Heisler and Jönsson (2006) showed that the localised expression of influx carrier AUX1 in L1 is likely required for auxin retention in the epidermis of the SAM, as suggested previously (Reinhardt *et al.*, 2003). They determined that the maintenance of a stable auxin pattern requires influx carriers, and AUX1 can specifically be redundant to other AUX1 family of proteins in stabilising the positioning of organs. In their study of *aux1* and *lax* mutants, Bainbridge *et al.* (2008) determined that auxin influx carriers AUX1 and LAX1 are likely redundant in function. This was confirmed using a variety of double, triple, and quadruple mutants for these genes. Quadruple mutants (*aux1*, *lax1*, *lax2*, and *lax3*) showed narrower or wider divergence angles in relation to wild-type meristems, indicating that these influx carrier genes, specifically AUX1 and LAX1, are important in stabilising auxin gradients to facilitate PIN1 polarisation during primordium initiation.

Sahlin *et al.* (2009) analysed the relationship between auxin-regulated polarised transport (efflux and influx) and the feedback mechanism in a broader sense.[3] Although the authors did not

---

[2] A complete list of parameter values is available in Heisler and Jönsson (2006).
[3] Sahlin *et al.* (2009) conducted a more general analysis of auxin transport based on the following general functional form of mediated auxin transport:

$$J_{a,i \to j} = p_a^H \left(\int_{a^H}^{cell} a_i - \int_{a^H}^{wall} a_{ij} \right) + p_{PIN} W_{ij} P_{ij} N(\Phi) h(a_i) - p_{AUX} W_{ij} A_{ij} N(\Phi) h_A(a_{ij}) \quad (13)$$

specifically analyse phyllotactic patterns, their two-dimensional model can generate various types of patterns: peaks, stripes, and re-entrant peaks. Concentration-based models provided new tools for generating different patterns in biology. These plausible models consider active biological mechanisms involved in organ development, and most of them are based on empirical data. However, the modelisation of auxin polarity transport remains unresolved. To address this problem, Fujita and Kawaguchi (2018) developed an auxin transport model in which they assumed the presence of a diffusible unknown molecule involved in auxin-PIN1 dynamics. This diffusible molecule would indicate the auxin concentration to neighbouring cells. They considered that the emergence of phyllotactic patterns requires an unknown molecular mechanism as well as auxin-PIN1 mutual interaction.

Auxin-Flux-based Polarisation Model of Stoma *et al.* (2008)

Stoma *et al.* (2008) developed a phyllotactic model based on the flux-based polarisation hypothesis. Cell–cell auxin transport in the SAM creates fluxes towards organ initiation sites. This hypothesis is

---

where $h(a_i)$ is an auxin-dependent function for mediated efflux and $h_A(a_{ij})$, the corresponding function for influx. They also included a new parameter (W_{ij}) which is the ratio between the area of the membrane in cell compartment i facing wall compartment ij and the volume of cell compartment i. This model also includes a protein cycling function and a production and degradation function that are similar to those of Heisler and Jönnson (2006).

By using this model, Sahlin *et al.* (2009) analysed feedback mechanisms allowing the emergence of patterns from a homogenous state. They concluded 'that the feedback has to be sensitive enough to differences in neighboring auxin concentration' (p. 68). Their analysis of the model shows that PIN1 is essential for the appearance of patterns; however, AUX1 does not appear to be required for pattern generation.

derived from Sachs' pioneering work (1969) proposing the canalisation hypothesis for the formation of vascular tissues. He assumed that the auxin transport polarity increases during vascular tissue induction. There is positive feedback between auxin concentration and its transporters, leading to the amplification of the flux and the formation of channels that will differentiate into vascular bundles. This canalisation hypothesis was used in mathematical models to analyse the pattern of vein formation (Mitchison, 1980, 1981; Feugier *et al.*, 2005; Rolland-Lagan and Prusinkiewicz, 2005; Fujita and Mochizuki, 2006). Stoma *et al.* (2008) extended the principles developed in these models to analyse phyllotactic patterns in the framework of auxin transport. Barbier de Reuille *et al.* (2006) also used a flux-polarisation-based model in their analysis of cell–cell signalling at the SAM in *Arabidopsis*. Through numerical simulations and experimental results, they also showed that the central zone of the SAM of *Arabidopsis* can accumulate auxin. They suggested that the central zone is important in the generation of auxin maxima at the site of incipient primordia.

The flux-based hypothesis differs from the *concentration-based* hypothesis used in the models of Smith *et al.* (2006b), Jönsson *et al.* (2006), and Heisler and Jönsson (2006). In these models, the relative auxin concentration in neighbouring cells differentially drives the polarisation and localisation of PIN1 in the membrane of adjoining cells, resulting in the movement (active export) of auxin against the auxin concentration gradient and leading to the creation of local maxima of auxin.

The flux-based hypothesis involves the establishment of a positive feedback loop in which a small flux between two cells reinforces itself through an increase in the local amount of PIN transporters, which will facilitate the evacuation of auxin in the flux direction. The flux-based polarisation model of Stoma *et al.* (2008) comprises two elements: auxin transport and flux-based polarisation.

Auxin transport

Stoma *et al.* (2008) assumed that the variation rate in auxin concentration results from three processes: (1) diffusion, (2) active auxin transport by PIN, and (3) local cell auxin synthesis and decay. Accordingly, they expressed the auxin variation in cell i as

$$\frac{\partial a_i}{\partial t} = -\frac{i}{V_i}\sum_{n\in N_i} S_{i,n}J^D_{i\to n} - \frac{i}{V_i}\sum_{n\in N_i} S_{i,n}J^A_{i\to n} + \alpha_a - \beta_a a_i \qquad (14)$$

where a_i is the auxin concentration in a cell i; $p_{i,n}$, the concentration of PIN proteins in the membrane between cells i and n; V_i, the volume of cell i; N_i, the set of neighbouring cells of cell i; $S_{i,n}$, the exchange surface between cells i and n; $J^D_{i\to n}$, the auxin flux due to diffusion from cell i to n; and $J^A_{i\to n}$, the active transport from cells i to n. The constant α_a is the rate of production of auxin in cells, and the constant β_α is the rate of degradation of auxin.

To describe passive diffusion between cells, Stoma *et al.* (2008) used Fick's first law, $J^D_{i\to n} = \gamma_D(a_i - a_n)$, where γ_D is the constant of permeability. It indicates the quantity of auxin that can move between cells per unit time. They modelled active transport between cells i and n based on Mitchison's formula (1980, 1981), $J^A_{i\to} = \gamma_\alpha(a_i p_{i,n} - a_n p_{n,i})$, where the constant γ_α represents the transport efficiency of the PIN pumps. Then, Eq. (21) becomes

$$\frac{\partial a_i}{\partial t} = -\frac{i}{V_i}\sum_{n\in N_i} S_{i,n}\gamma_D(a_i - a_n) - \frac{1}{V_i}\sum_{n\in N_i} S_{i,n}\gamma_A(a_i p_{i,n} - a_n p_{n,i}) + \alpha_a - \beta_\alpha a_i \qquad (15)$$

Flux-based polarisation

Stoma *et al.* (2008, p. 4) assumed that 'the concentration of PIN proteins $p_{i,n}$ in cell i transporting auxin to cell n changes due to: (i) insertion in the membrane induced by the flux and (ii) background insertion and removal of PIN from the membrane'. The net flux of auxin across the membrane from cells i to n is $J_{i \to n} = J_{i \to n}^{D} + J_{i \to n}^{A}$. Then, the PIN variation rate across the membrane is

$$\frac{\partial p_{i,n}}{\partial t} = \Phi\left(J_{i \to n}\right) + \alpha_p - \beta_p p_{i,n}, \tag{16}$$

where Φ is a function that defines the intensity of PIN insertion into the membrane owing to the feedback of auxin flux. Stoma *et al.* (2008) noted that the nature of this function will determine the type of canalisation process.[4] The constant $\propto_a$ is the rate of background insertion in the membrane and the constant β_p, the background removal rate from the membrane.

The numerical solutions of the equations describing flux-based polarisation were used to generate a spiral phyllotactic pattern (average divergence angle: 137°) on a virtual growing meristem. Two parameters play an essential role in this model: the constant (κ) controlling the feedback level of auxin fluxes on PIN insertion in the membrane and the sink (primordium) initiation threshold. An increase in κ extends the intensity of inhibitory fields around sinks (primordia). When a particular auxin threshold ω (sink initiation

---

[4] Stoma *et al.* (2009) used two types of functions: a linear function $\Phi_L = \left(\mathfrak{J}_{i \to n}\right) = \kappa\left(\mathfrak{J}_{i \to n}/\mathfrak{J}_{ref}\right)$ and a quadratic function $\Phi_Q\left(\mathfrak{J}_{i \to n}\right) = \kappa\left(\mathfrak{J}_{i \to n}/\mathfrak{J}_{ref}\right)^2$, where κ is a constant that controls the feedback level of auxin fluxes on PIN insertion in the membrane and $\mathfrak{J}_{ref}$, an arbitrary reference flux that enables keeping constant units in the equation. If no additional PIN is inserted in the membrane, more auxin molecules come in than molecules go out. In this case, $\Phi = 0$.

threshold) is reached, the amount of hormone is sufficient to initiate a new sink (primordium) at the farthest distance from the preceding two sinks. Stoma (2008) also showed that the frequency of primordium initiation increases with a corresponding decrease in the initiation threshold ω.

A primordium, therefore, acts as a sink which results in auxin depletion in its immediate surroundings. In turn, this prevents the formation of new primordia near existing ones (Fig. 4.8). During later stages of development and tissue growth, new space becomes available and conditions are created for a new primordium that will

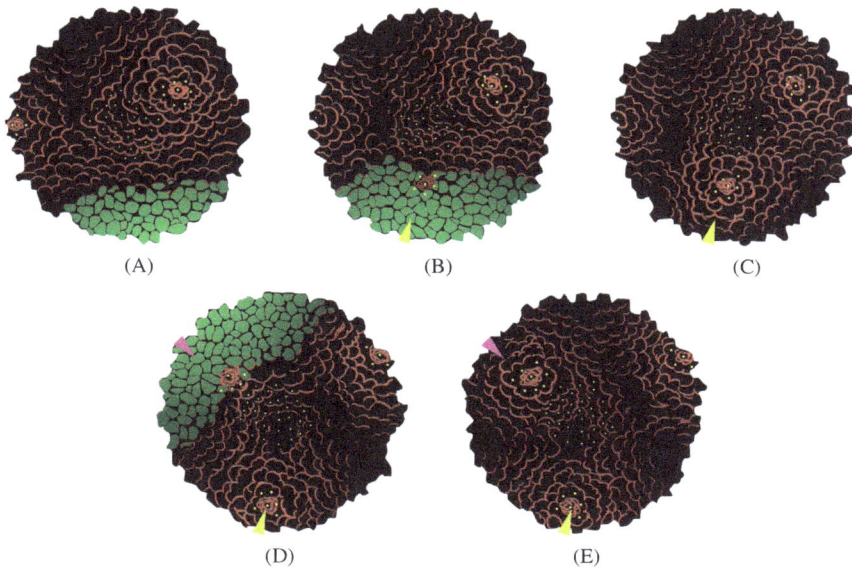

Fig. 4.8. Dynamic model. (A–E) 2D top view of a virtual meristem following the flux-based model showing the dynamic of PIN distribution, auxin concentrations, and primordia initiation. The sequence shows the initiation of two consecutive primordia (arrowheads). The sequence starts with auxin accumulation in the zone that is the farthest away from the existing primordial. When a threshold is reached, the maximum acquires primordium identity and becomes a sink. (From Stroma *et al.*, 2008.)

become a sink to be initiated. Van Berkel *et al.* (2013) analysed current models of polar auxin transport in the SAM in detail and concluded that neither concentration- nor flux-based models can fully explain self-organising patterns; they likely require additional assumptions at the spatial 'tissue' level. This may imply the following: (1) auxin patterning is not completely self-organised and requires a tissue-specific response or (2) current feedback mechanisms need to be revised to properly explain self-organised auxin patterning.

Although these two types of models are different, both can generate the same phyllotactic patterns. In fact, the appearance of phyllotactic patterns in both is based on the transposition of the auxin influx in terms of concentration threshold in relation to the size of the apex. Thus, in phyllotactic modelling, the emergence of phyllotactic patterns is independent of the mechanism leading to auxin accumulation in different zones of the SAM.

Floral Phyllotaxis

Many phyllotactic principles and models that have already been discussed for shoot systems can apply to the initiation of flowers and floral organs. Similarly, the role of auxin in inducing the initiation of floral organs and the establishment of meristem-organ boundaries are well-documented in the elaboration of 'floral ground plans' (Smyth, 2018). For example, there is growing evidence that auxin plays a key role in pattern formation in capitula as well as floral meristem identity (Zoulias *et al.*, 2019). This particular section highlights a few recent models dealing specifically with floral phyllotaxis.

Kitazawa and Fujimoto (2015) recently developed a dynamic model dealing specifically with the arrangement and number of floral organs. Their model assumes two developmental processes: sequential and periodical initiation of primordia following the Hofmeister rule and mutual repulsion among primordia during SAM growth. Based on the inhibitory field model (e.g. Smith *et al.*, 2006b) and the contact pressure model (e.g. Hellwig *et al.*, 2006), they developed a theoretical model comprising two parts, the initiation process and the growth process. These respectively correspond to two successive interactions between emerging primordia: (1) a new primordium is formed where the inhibition coming from pre-existing primordia is the lowest, and (2) a mutual repulsion between primordia causes the displacement of recently initiated primordia around the floral meristem (Fig. 4.9). By using this model, they reproduced the phyllotactic organisation of spiral and whorled patterns (i.e. tetramerous and pentamerous flowers).

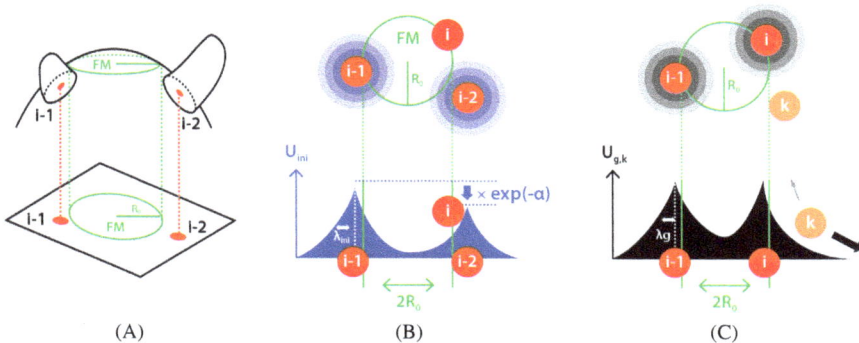

(A)   (B)   (C)

Fig. 4.9.   Emergence of multiple whorls in model simulations. (A) Geometric assumptions of the model. (B) The initiation process. A new primordium (i) is initiated at the edge of the floral meristem (FM; green circle) where the initiation potential U_{ini} takes the minimum value. i, $i − 1$, and $i − 2$ are the primordium indices that denote the initiation order. U_{ini} exponentially decreases with time (alpha) and the distance between primordia (λ_{ini}). (C) The growth process. Each primordium (k) moves at the outside of the circular FM depending on the growth potential $U_{g,k}$. Primordium k rarely moves against the gradient (grey thin arrow) but mostly follows the gradient (black thick arrow). (From Kitazawa and Fujimoto, 2015.)

The initiation process is based on an inhibition potential function as is the case in Douady and Couder's model (1996a). The meristem is represented by a circular disk of radius R_0 and primordia are represented by points with polar coordinates (R_0, θ), where θ is the divergence angle. A new primordium is formed at the edge of the disk when the value of the inhibition potential function U_{ini} by pre-existing primordia is at a minimum.

The potential at the initiation of the ith primordium is given by

$$U_{ini}(\theta) = \sum_{j=1}^{i-1} \exp\left(-\alpha\left(i-j-1\right)\right)\exp\left(-\frac{d_{ij}}{\lambda_{ini}}\right) \qquad (17)$$

In this equation, the inhibition potential decreases exponentially with the age of a primordium and the distance between a new primordium and pre-existing primordia with decay rate α. The exponential function $\exp\left(-\frac{d_{ij}}{\lambda_{ini}}\right)$ is a function of θ, where λ_{ini} is the duration of decay. d_{ij} corresponds to the distance between a new primordium i and a pre-existing primordium j at (r_j, θ_j) so that

$$d_{ij} = \sqrt{R_0^2 + r_j^2 - 2R_0 r_j \cos(\theta - \theta_j)}. \qquad (18)$$

For the growth process, Kitazawa and Fujimoto (2015) assumed that all primordia repel each other after their initiation. The repulsion exerted on the kth primordium corresponds to another exponentially decaying potential. For a system with i primordia ($1 \le k \le i$),

$$U_{g,k}(r,\emptyset) = \sum_{j=1,j\neq k}^{i} \exp\left(-\frac{d_{kj}}{\lambda_g}\right), \qquad (19)$$

where the duration of decay λ_g can differ from λ_{ini}. The primordia shift to the periphery of the floral meristem to find a location where repulsion is weaker. This formulation is similar to the contact pressure model of Ridley (1982) and Hellwig *et al.* (2006), where the divergence angle is re-corrected after the initiation of primordia.

For the numerical simulation, a new primordium is introduced where the U_{ini} value is at a minimum. Then, a primordium $k(r_k, \theta_k)$ selected randomly among the existing primordia shifts outside the meristematic zone $(r'_k \geq R_0)$ to a new position (r'_k, θ'_k), where r'_k and θ'_k are chosen randomly. If the growth potential of the new position $[U_{g,k}(r'_k, \theta'_k)]$ is lower than that of the previous one $[U_{g,k}(r_k, \theta_k)]$, then the primordium will move to its new position; otherwise, it will move with a probability corresponding to $P_{mp} = \exp(-\beta \Delta U_g)$, where $\Delta U_g = U_{g,k}(r'_k, \theta'_k) - U_{g,k}(r_k, \theta_k)$ and β is a parameter for stochasticity. If $P_{mp} = 0$, there is no growth potential, and the primordium will stay at the edge of the meristem $(r = R_0)$ until the next one arises; however, if $P_{mp} \neq 0$, the primordium can shift randomly outside the edge of the meristem.

Based on these equations and rules, Kitazawa and Fujimoto (2015) showed that several self-organised whorls can appear after the sequential initiation of primordia at the proper meristem size as is observed in many flowers. The whorled pattern of organisation is due to an arrest in radial displacement and the mutual repulsion between primordia throughout the growth process. The meristem size is the parameter (similar to Γ in other models) that controls the transition between different phyllotactic patterns. A small value $(R_0 < 10)$ is linked to a non-whorled organisation, and whorled patterns emerge at higher values $(R_0 \approx 20)$. By using this model, they produced tetramerous and pentamerous whorled patterns that are common among flowering plants. They also showed a quantitative correlation between their model and measures taken on the pentamerous flower of *Silene coeli-rosa*. Kitazawa and Fujimoto (2018) were able to highlight the existence of developmental genetic constraints in the floral phyllotaxis of *Anemone*.

Van Mourik *et al.* (2012) studied the role of auxin in the formation of floral organs in *Arabidopsis* from empirical and theoretical viewpoints. They combined a cellular growth model of the floral meristem with a polar auxin transport model to simulate the initiation of whorled organ patterning. They based their model of the developing floral meristem on the auxin transport models developed for phyllotaxis in shoots (Barbier de Reuille *et al.*, 2006, Jönsson *et al.*, 2006, Smith *et al.*, 2006a), particularly that of Jönsson *et al.* (2006).

Van Mourik *et al.* (2012) experimentally showed that auxin accumulation sites correspond to the location where floral primordia are initiated. In their theoretical model, the localisation of auxin accumulation peaks is a function of the growth rate and auxin transport parameters. This enabled them to predict the positioning and timing of floral organ initiation. They also considered floral mutations to test their model by adjusting the model parameters accordingly. They concluded that auxin transport is sufficient to explain phyllotactic patterns in mutant flowers based on the similarities observed between predictions from the model and actual mutants. Their model is based on the idea that auxin is pumped through the cell wall by PIN efflux carriers located at the cell membrane. However, as auxin inhibits endocytosis of PIN's molecules in neighbouring cells, the auxin flux is directed towards cells with a higher concentration than their neighbours. They also assumed passive transport through a cell-to-cell diffusion process that is not mediated by PIN molecules. Passive transport tends to attenuate the difference in auxin concentration between cells, whereas active transport increases it. An increase in the contribution of active transport leads to a decrease in distance between auxin concentration peaks.

Van Mourik *et al.* (2012) made two assumptions regarding the growth rate of meristematic cells: (1) the growth rate of a cell increases proportionally to its auxin concentration and (2) when a cell reaches a threshold auxin concentration, it differentiates into a particular organ cell and the auxin is pumped from the L1 layer as is the case in the models of Smith *et al.* (2006a) and Barbier de Reuille *et al.* (2006).

Van Mourik *et al.* (2012) restricted their simulations to the part of the meristem where the primordia of floral organs are initiated. They used a circular planar approximation of the meristem. The equations for auxin concentration (A_i) and PIN concentration (P_i, P_{ij}) are

$$\frac{dA_i}{dt} = T \sum_{j \in N_i} \left(P_{ji} f_1 \left(A_j \right) - P_{ij} f_1 \left(A_i \right) \right)$$

$$+ D \sum_{j \in N_i} L_{ij} \left(A_j - A_i \right) + p - d_i A_i + \phi_b L_{ij} \tag{20}$$

$$\frac{dP_i}{dt} = \sum_{j \in N_i} \left(-k_1 f_2 \left(P_i \right) f_3 \left(A_j \right) + k_2 P_{ij} \right) \tag{21}$$

$$\frac{dP_{ij}}{dt} = k_1 f_2 \left(P_i f_3 \left(A_j \right) - k_2 P_{ij} \right). \tag{22}$$

The first equation describes the auxin concentration (A_i) in a cell i, which depends on P_{ij}, the amount of PIN molecules in cell i at the wall facing neighbour j. The second and third equations describe the PIN molecule cycle between the interior of the cell (endosome) (P_i) and the wall (P_{ij}). Van Mourik *et al.* (2012) defined

$j \in N_i$ as the set of cell j adjacent to cell i. The constants T and D are respectively the active and passive transport coefficients, L_{ij} is the cell surface, and p is the constant local auxin production. d_i accounts for both decay d_{dec} and deletion d_{dep} after primordium initiation; for $A_i > 0$, $d_i = d_{dec} + d_{dep}$. k_1 and k_2 characterise the PIN cycling between the interior of the cell (endosome) and the cell wall, and f_1 is a Michaelis-Menten saturation function. The auxin flux at the periphery of the theoretical planar meristem is represented by the parameter ϕ_b which reflects the auxin fluxes through the L1 layer towards the apex of the floral meristem. The increase in the cell surface is described by the function $\frac{ds_i}{dt} = \left(1 + \frac{k_4 A_i}{k_5 + A_i}\right)\gamma$, where s is the surface area; k_4, the auxin-dependent growth rate; k_5, the half-maximum value; and γ, the growth rate constant. Thus, cell growth depends on the following factors: internal turgor pressure, the strain of cell wall, and pressure exerted by surrounding cells. It is determined by minimising the generalised potential energy (H) equation:

$$H = \lambda_A \sum_i \left(s(i) - S_T(i)\right)^2 + \lambda_m \sum_j \left(l(j) - L_T(j)\right)^2 \qquad (23)$$

where λ_A is the turgor pressure; λ_m, a spring constant; $s(i)$, the actual cell area; $S_T(i)$ the predefined resting area; $l(j)$, the actual wall length; and $L_T(j)$, the predefined resting length. The summation is performed over cells (i) and polygon edges (j) of the cells. The minimum value of Eq. (23) is determined by using a Monte Carlo algorithm that can generate different random deformations of the typical cell shape. This introduces a random component in each simulation (see the original paper for a more detailed description of the equations and the algorithm).

Van Mourik *et al.* (2012) simulated the position of four accumulation sites corresponding to the sepals on *Arabidopsis*

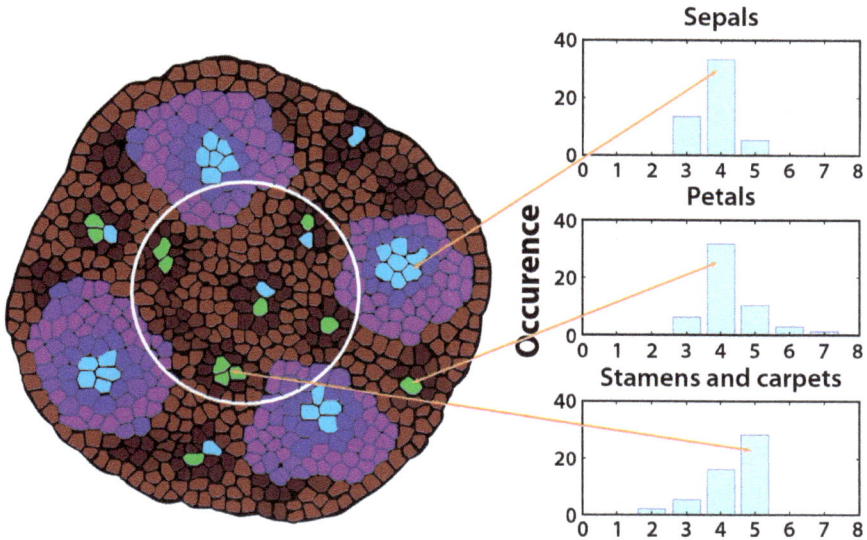

Fig. 4.10.   Bar graph of the emerging organ numbers for 50 simulations. The arrows indicate the organs on the simulated meristem. The circle separates the areas of the two outer whorls and two inner whorls. (From Van Mourik *et al.*, 2002.)

flowers (Fig. 4.10). A comparison of the position of the auxin peaks with the position of the sepals on a real floral meristem revealed a good correlation between the predictions and the empirical data. However, other floral organs show greater variation in the simulations; in these cases, they observed slightly more than four auxin maxima for the petals and less than six and two for stamens and carpels, respectively. This variation is partly explained by the use of a 2D approximation, instead of a 3D one, of the floral meristem.

The authors also tested their model's predictive value by using weak (moderate change in organ number) and strong (large change in organ number) mutants. They adjusted the model parameters according to the floral phenotype of the mutants and found good agreement between simulated and observed morphologies. They concluded that sepals constrain the future position of the smaller auxin maxima associated with petals, stamens, and carpels and that polarisation of auxin transport mechanisms could be involved in floral organ patterning as is the case for phyllotaxis in shoots.

Conclusion

Recent studies have pointed out the multiple roles of several genes in the generation of phyllotactic patterns. Can these specific gene effects be understood in a broader, system-wide context? At the heart of this system are highly organised meristematic centres such as the shoot apical meristem and the inflorescence and floral meristems. Each of these stem-cell rich centres constitutes the framework for the generation of form but imposes constraints for the development of these forms. Within the framework of meristematic centres and beyond, the activity and action of genes can take place at different hierarchical levels (i.e. within cells, between cells, within tissues, as well as within and between organs). As outlined in this chapter, studies using mutants highlighted several genes operating at the level of meristem initiation, organisation, and maintenance, but also, and more specifically, at the level of organ identity and site of initiation (including number of organs and their position). These multiple levels of action and interaction indicate that none of the specific mutations identified to date directly affects phyllotaxis, although a general pattern of alteration of growth modalities or processes seems to be emerging. These alterations include those in: growth rate, duration, and distribution; symmetry; branching; inter-primordial growth; ontogenetic displacement; and fusion. Developing (dynamic) structures can, therefore, be defined by a combination of morphogenetic processes operating at different levels of an organisation, from individual cells to whole organs (Lacroix *et al.*, 2005). These processes operate within the constraints of the SAM and can affect positional information as well as size relationships between primordia and the meristem, among other relationships, ultimately affecting phyllotactic patterns.

The role of auxin as an effector of pattern formation and versatile morphogen (in terms of its potential to affect growth) is now well-established and offers a solid foundation for further exploring and refining our understanding of phyllotaxis. Many recent studies on phyllotaxis also highlighted the importance of combining theoretical models and empirically testing basic assumptions related to the initiation of organs and the value of parameters (Kramer, 2008). Another important aspect that is emerging from biological and molecular studies on phyllotaxis is the role of mechanical signals and their function in terms of feedback on morphogenesis. This particular topic is the subject of Chapter 5.

Appendix 1

Examples of gene mutations associated with phyllotactic parameters and with apex and leaf sizes and forms. R, plastochrone ratio; b, the ratio between the diameter of the element and that of the apex; m, n, number of contact parastichies; d, angle of divergence; γ, angle between parastichies. X indicates the association found in studies conducted so far.

Gene name (**abbreviation**); Plant species (common name)	Phyllotactic parameters					Apex		Leaf		Authors: Gene function
	d	m, n	R	b	γ	Size	Form	Size	Form	
Abnormal phyllotaxy1 (**abphyl1**); *Arabidopsis thaliana* (Arabidopsis)	X	X		X		X	X	X		Jackson and Hake (1999), Braybrook and Kuhlemeier (2010), Bar and Ori (2014): Position of leaves and meristem size
Aintegumenta (**ant**); *A. thaliana*									X	Sassi and Vernoux (2013): Organ growth and cell proliferation
Arabidopsis histidine phosphotransfer protein6 (**ahp6**); *A. thaliana*	X		X							Besnard *et al.* (2014a,b): Regulator of the plastochrone at the meristem
Bellringer (**bell**); *A. thaliana*	X	X	X							Byrne *et al.* (2003): Disruption of pattern (more frequent initiation of organs)

Gene							References
Clavata3 (**clv3**); *A. thaliana*	X			X			Reinhardt (2005a), Sassi and Vernoux (2013): Active in central zone (stem-cells)
Decussate (**dec**); *Oryza sativa* (Rice)		X	X	X			Itoh *et al.* (2012): Distichous to decussate phyllotactic change; size of apex
Cup-shaped cotyledon 2 and 3 (**cuc2, cuc3**); *A. thaliana*		X	X	X	X		Burian *et al.* (2015): Double mutant, decreased angular width of primordia; post-meristematic effects
Fasciata (**fas**); *A. thaliana*	X	X		X			Kaya *et al.* (2001): Cellular and functional organisation of SAM; mutant: broader and flatter SAM
Forever young (**fey**); *A. thaliana*			X			X	Callos *et al.* (1994): Regulation of meristem development
Hairy meristem (**ham**); *Petunia* sp. (Petunia)	X			X			Stuurman *et al.* (2002): Maintenance of meristem

(Continued)

(Continued)

Gene name (**abbreviation**); Plant species (common name)	Phyllotactic parameters					Apex		Leaf		Authors: Gene function
	d	m, n	R	b	γ	Size	Form	Size	Form	
Katanin1 (**ktn1**); A. thaliana	X									Jackson et al. (2019): Cell shape heterogeneity
Leafy (**lfy**); A. thaliana	X									Weigel et al. (1992), Sassi and Vernoux (2013): Meristem identity (vegetative or floral)
Lycopersicon esculentum expansin18 (**LeExp18**); Lycopersicon esculentum (Tomato)	X									Reinhardt et al. (1998), Kwiatkowska (2004): Position of leaves
Monopteros (**mp**); A. thaliana									X	Reinhardt (2005a): Organogenesis
Nicotiana tabacum homeobox1 (**nth1**); Nicotiana tabacum (Tobacco)			X						X	Tamaoki et al. (1999): Leaf shape, plastochrone ratio

Oriza sativa pinhead1 (**OsPNH1**); *O. sativa*	X				Nishimura *et al.* (2002): Shape of meristem, site of leaf initiation
Phantastica (**phan**); *Antirrhinum* sp. (Snapdragon)			X	X	Waites *et al.* (1998): Growth and dorsiventrality of lateral organs; SAM form
Pin-formed1-1 (**pin1**); *A. thaliana*	X		X		Reinhardt *et al.* (2000, 2005), Smith *et al.* (2006b): Position of leaves
Plethora (**plt**); *A. thaliana*	X				Prasad *et al.* (2011), Palauqui and Laufs (2011): Shoot organ positioning Pinon *et al.* (2013): Regulation of PIN1 activity; tuning mechanism of polar auxin transport; Activity in meristem centre is necessary for typical phyllotaxis

(Continued)

(Continued)

Gene name (*abbreviation*); Plant species (common name)	Phyllotactic parameters					Apex		Leaf		Authors: Gene function
	d	*m, n*	*R*	*b*	*γ*	Size	Form	Size	Form	
Revoluta (**rev**); *A. thaliana*	X									Otsuga *et al.* (2001): Position of lateral meristems
Rough sheath2 (**rs2**); *Zea mays* (Maize)									X	Schneeberger *et al.* (1998): Leaf form (*Knox* gene)
Shoot organisation (**sho**); *O. sativa*	X	X	X	X	X		X	X	X	Itoh *et al.* (2000): Position of leaves; SAM structure
Shoot meristemless (**stm**); *A. thaliana*	X									Long and Barton (2000), Kwiatkowska (2004): Meristem regulation; maintenance of indeterminate character of cells (*Knox* gene)
Terminal ear 1 (**te1**); *Z. mays*	X			X					X	Veit *et al.* (1998): Regulation of leaf initiation

Gene											Reference/Function
Ultrapetala (**ult**); A. thaliana										X	Reinhardt (2005a): Mutant, increase in number of organs and meristem
Wuschel (**wus**); A. thaliana										X	Sassi and Vernoux (2013): Mutant, no SAM
Yucca (**yuc**); A. thaliana	X										Sassi and Vernoux (2013): Lateral organ initiation and development (initiation site)
Yabby (**yab**), kanadi (**kan**), Homeodomain-leucine zipper (**hd-zipIII**); A. thaliana									X		Sassi and Vernoux (2013): Dorsiventrality of leaves

5 Biophysical Aspects of Phyllotaxis

Studies have long investigated the interpretation and modelling of phyllotactic patterns in terms of physical processes, particularly Airy's (1873), van Iterson's (1907), Schwendener's (1878, 1909), and, more recently, Adler's (1974) geometrical models of contact pressure. Schwendener's book entitled 'Mechanical theory of leaf position' (1878) presents a biophysical interpretation of phyllotaxis. Later, Church (1904, p. 229) speculated on the extent to which phyllotactic diagrams may also be '...taken as the expression of a field of distribution of growth energy, comparable, for example, to manifestations of distribution of the physical energy of the electromagnetic field'. Snow and Snow (1951) discussed the problem of tension in the apex with regard to leaf initiation and questioned the presence of tension or pressure in the superficial layers of the apices of flowering plants. After the publication of his fundamental phyllotactic study in 1907, van Iterson published 'New studies in phyllotaxis' (1964) in which he considered the apex of plants as analogous to a vibrating 'uniformly stretched elastic membrane', where 'The foregoing then leads to the hypothesis (to be elaborated in a later study) that the apical cones of higher plants (when forming their leaf primordia) behave like small *resonators...*' (p. 142). He noted that this would explain the dominance of certain common phyllotactic patterns in nature.

In the last decade, new models dealing with physical processes acting at the SAM level have provided novel interpretations for the emergence of phyllotactic patterns (Newell and Pennybacker, 2013; Pennybacker *et al.*, 2015). Shipman and Newell (2005, p. 192) stated that the manifestation of buckling, which is initially due to uniform compressive stress (although the stress could be anisotropic and become nonuniform after buckling), constitutes the initial stage of pattern formation. This interpretation is primarily based on the pioneering studies of Green and collaborators.

Following a series of experimental studies, Green (1992a,b) proposed that the tunica or 'outer skin' consisting of 1–2 cell layers in meristems shows minimal energy buckling behaviour and that the underlying corpus supplies upward pressure. Many other studies also indicated that mechanical processes could be involved in the formation of phyllotactic patterns. For example, Hernández and Green (1993) showed that mechanical external constraint can change the identity of bracts in sunflower capitula. Similarly, Green (1999) induced the initiation of a new row of leaves on the SAM by applying a mechanical constraint with a glass film. Dumais and Steele (2000) revealed the presence of circumferential compression near the generative region of sunflower capitula by using microsurgical manipulations such as cuts and physical constraints. Steele (2000) noted a quantitative relationship between the size of primordia and the thickness of the tunica. He postulated that a turgor pressure of 7–10 atmospheres in plant cells could play a role in phyllotactic processes. The application of expansin protein on the SAM of tomato plants also indicated a possible role for turgor and physical forces (Fleming *et al.*, 1997). Different studies using modelling and direct measurements of the mechanical properties of the meristem demonstrated that the central zone (CZ) of the SAM is stiffer than the peripheral zone (PZ) (Milani *et al.*, 2011, 2014; Reinhardt *et al.*, 2012). At the cellular level, the modification of the structural components of the cell wall affects phyllotactic organisation (Nakayama and Kuhlemeier, 2008; Peaucelle *et al.*, 2008). This indicates that physical modification of the cell wall can influence physical processes acting at the macroscopic level. Auxin reduces tissue rigidity before the outgrowth of organs at the SAM in *Arabidopsis*, and de-methyl-esterification of pectin is necessary for this reduction to occur (Braybrook and Peaucelle, 2013). Robinson *et al.* (2013) provided an interesting overview of mechanical dynamics and their influence on cellular and developmental processes in the SAM.

Green's central hypothesis (1992a, 1999) is that compressive physical stresses acting at the SAM level lead to the buckling of the tunica and ultimately explain primordium initiation and the emergence of phyllotactic patterns. This compression could result from two processes linked to the growth mode (Dumais and Steele, 2000; Shipman and Newell, 2005): differential growth of the tunica, as suggested by Schüpp (1914), Priestly (1928), and Green (1992a,b), or the pressure exerted on the tunica by the faster growth of the corpus, as suggested by Selker *et al.* (1992), Steucek *et al.* (1992), and Steele (2000), who viewed the tunica as a pressurised shell. However, Shipman and Newell (2005) noted that if the tunica is represented by a spherical shell, by analogy to the SAM geometry, the tunica will experience a tensile stress rather than a compressive stress inducing buckling.

Based on the principles elaborated in their biological model, Green *et al.* (1998) and Steele (2000) simulated the propagation of spiral and whorled phyllotactic patterns. Recently, Shipman and Newell (2005) extended this model considerably to include different phyllotactic patterns and the planform to which they are associated. Here, *phyllotactic planform* refers to both the shape and the arrangement of elements. For example, in *Cactaceae*, the tilling of tubercle-like structures (corresponding morphologically to

Fig. 5.1.   Common planforms on plants are (**A**) ridges, (**B**) irregular hexagons, (**C**) parallelograms, and (**D**) staircase parallelograms. (**E**)–(**H**) Theoretical reproductions of each planform based on the mechanical model.(From Shipman and Newell, 2005; reprinted with permission from Elsevier.)

the leaf base) on the stem surface translates to different geometrical figures depending of the species: ridges, hexagonal, rhombic, and offset-rhombic (see Fig. 5.1). The phyllotactic patterns corresponding to these different morphologies were described anatomically by Bilhuber (1933) and analysed more recently from a biomechanical perspective by Newell *et al.* (2012). Contrary to the model of Green *et al.* (1998), Shipman and Newell's analysis (2005) addresses the question of the shape of primordia (i.e. ridges, hexagons, parallelograms) and the transitions between phyllotactic patterns. The observed planforms and phyllotactic patterns which minimise the elastic energy of the annular region of initiation of primordia are self-emergent and are independent of boundary conditions.

Shipman and Newell (2004) demonstrated how common naturally occurring phyllotactic patterns can be generated by the energy-minimising buckling pattern appearing in a compressed shell (corresponding to the plant's tunica) on an elastic foundation.

They demonstrated how minimising the elastic energy on a surface analogous to the tunica of plants can generate spiral or whorled patterns and transitions between them. As in many theoretical models, Fibonacci-like sequences and the golden angle are natural manifestations of the parameters involved in this model.

Representation of SAM in Biomechanical Models

In the biomechanical model of Newell *et al.* (2008a), the theoretical SAM consists of four zones: annular zones 1, 2, and 3 belong to the tunica and zone 4 corresponds to the corpus which constitutes an elastic foundation (Fig. 5.2). Zone 1 grows slowly in relation to zone 2, which the authors called the *generative region* because it is where primordia are initiated as bumps. This region experiences compressive stress owing to the overall growth mode of the SAM. The position of primordia in the generative zone ultimately determines the phyllotactic pattern.

The generative region (zone 2), represented as a thin elastic shell of radius R and thickness h (tunica), is attached or continuous with an elastic foundation, the corpus (Fig. 5.2). Two other geometrical parameters characterise the SAM: radial curvature R_r and circumferential curvature R_a. This external boundary of zone 2 is characterised by a hardening tunica which was previously subjected to compressive stresses as a result of growth pressure and a progressive hardening process. This leads to the formation of primordial bumps arranged in regular patterns on the SAM surface. As the SAM develops and as new primordia are initiated, existing primordia transition to the hardening area of the apex.

In their model, Shipman and Newell (2005) recognised three possible geometric patterns of the SAM which they felt

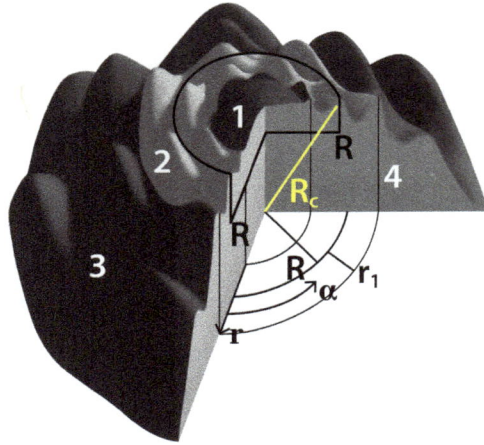

Fig. 5.2.   Schematic diagram of SAM. The SAM consists of a thin skin (the tunica, regions 1, 2, and 3 in the diagram) attached to a foundation of less well-organised cells (the corpus, 4). Cells in region 1 show little growth activity, and region 2 is the generative region in which some active cell division occurs and new primordia are first seen to form. As the plant tip grows, cells move radially outwards in the reference frame of the diagram's north pole from regions 2 to 3. In Region 3, primordia further develop into mature phyla such as leaves. We define radial r and angular a coordinates on the generative region (region 2) by projecting the surface to polar coordinates on the plane. (From Newell *et al.*, 2008a; reprinted with permission from Elsevier.)

corresponded to those observed in plants. However, they mentioned that no comprehensive study has reported on the presence of these geometries. Such a study would be very useful given that nearly all phyllotactic observations and phyllotactic models are based globally on a spherical apical geometry. In the first geometrical pattern (Fig. 5.3A), the annular generative region (white band) is located on a sphere where the corpus constitutes the inner part. Surprisingly, they stated that this type of geometry is not the typical shape of the generative region. Nevertheless, if one considers studies on plant development, this type of geometry characterises most plants. In the second type of geometry

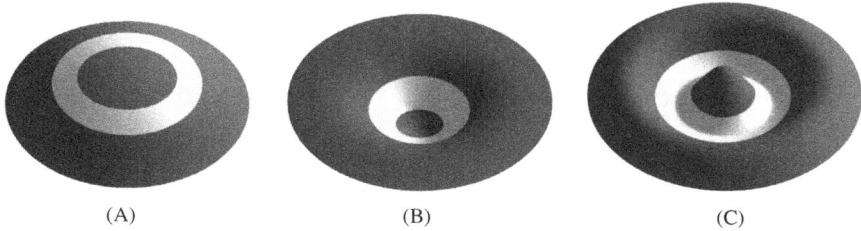

(A)   (B)   (C)

Fig. 5.3. **(A)** Spherical $(R_\alpha, R_r > 0)$, **(B)** inverted spherical $(R_\alpha, R_r < 0)$, and **(C)** hyperbolic $(R_\alpha > 0, R_r < 0)$ geometries of the generative region, denoted by light shading. (From Shipman and Newell, 2005; reprinted with permission from Elsevier.)

(Fig. 5.3B), the generative region is also part of a sphere; however, the inner corpus is subadjacent to the external outermost portion of the sphere. The third type (Fig. 5.3C) is called hyperbolic geometry and is characterised by a generative region surrounding a central dome. They stated that the geometrical patterns of the second and third types occur in cacti and sunflower capitula.

The three abovementioned types of geometry are described by using the *signed radii of curvature* which gives the direction and magnitude of the curvature of the shell. The position on any point of the shell is given by using the radial r and angular α coordinates of the shell projection onto a plane. Then, R_r and R_α represent the radii of curvature along the r and α coordinate lines, respectively. Two positive radii of curvature $(R_r, R_\alpha > 0)$ characterise the spherical geometry (Fig. 5.3A). It also follows that two negative radii of curvature $(R_r, R_\alpha < 0)$ describe an inverted sphere (Fig. 5.3B). Additionally, $R_r < 0$ and $R_\alpha > 0$ define a hyperbolic geometry of the SAM (Fig. 5.3C).

Shipman and Newell's model is aimed at determining the buckling configurations that minimise elastic energy. They defined elastic energy as an integral over M with contributions from different factors including bending energy, strain energy, in-surface deformation, and an external potential containing the influence of the elastic foundation, radial pressure, and growth in the tunica. Shipman and Newell (2004) found that Fibonacci-like sequences and hexagonal configurations minimise the elastic energy. They also reproduced a variety of naturally occurring phyllotactic patterns.

Periodic Generation of Phyllotactic Patterns

In their model, Shipman and Newell (2005) considered the deformation w of the generative region instead of considering the formation of a specific primordium. Their first concept is that phyllotactic patterns corresponding to the deformations $w(r, \alpha)$ of the generative region on the SAM surface (where r and α are respectively radial and angular coordinates from the centre of the SAM) are generated by linear combinations of elementary periodic functions (cosines) in the form

$$w(r,\alpha) = \sum_{j=1}^{N} a_j(l_j, m_j)\cos(l_j r + m_j \alpha) \tag{1}$$

where a is the amplitude; l, the radial wavenumber; and m, the circumferential wavenumber. The summation comprises all vectors that constitute the active set. The set in question represents all shapes that are amplified or slightly dampened. In this model, the choices of amplitude a_j and wavevectors $\vec{k}_j = (l_j, m_j)$ are those that minimise the elastic energy of the generative region under mechanical stress due to growth.

The active set includes wavevector triplets

$$\vec{k}_m = (l_m, m), \vec{k}_n = (l_n, n), \vec{k}_{m+n} = \vec{k}_m + \vec{k}_n = (l_m + l_n, m + n)$$

which strongly reinforce each other and minimise energy. The interaction between wavevectors is responsible for the appearance of phyllotactic patterns on theoretical SAMs. The magnitude of the quadratic interactions is directly proportional to C, the curvature of the surface of the tunica before buckling. Shipman and Newell (2005) assumed that there is a periodic formation of the same pattern in the generative region. This pattern is displaced away from the SAM centre and is replaced by the same pattern over time at regular intervals. To obtain a physical representation of the pattern, the function $w\,(r,\,\alpha)$ must be plotted over different periods. In this function, $s = r$ or $s = \ln(r)$ depending on whether the movement of the primordium in relation to the SAM centre is constant or exponential. If the analysis concerns only the generative region, the authors used r; however, if the analysis considers the pattern beyond the generative region, they used s. However, in their Eq. (1), it is necessary to empirically determine the wavevectors $\vec{k}_j = (l_j, m_j)$ and real amplitudes a_j, where l and m are real numbers. For this purpose, they used a physical equation minimising the elastic energy of the curved generative region under mechanical stress due to growth. Given that this model is based on the interaction of spatially periodic ripples, the phyllotactic lattice is characterised in terms of vectors.

Geometrical Representation and Vectorial Representation

Shipman and Newell (2005) noted a difference between a dual pair of lattice generators (w) used in geometrical representations of phyllotactic patterns and the pair of wavevectors used in their model. Only the pair of wavevectors can describe the shape of the primordia. For example, in (Fig. 5.4A), the

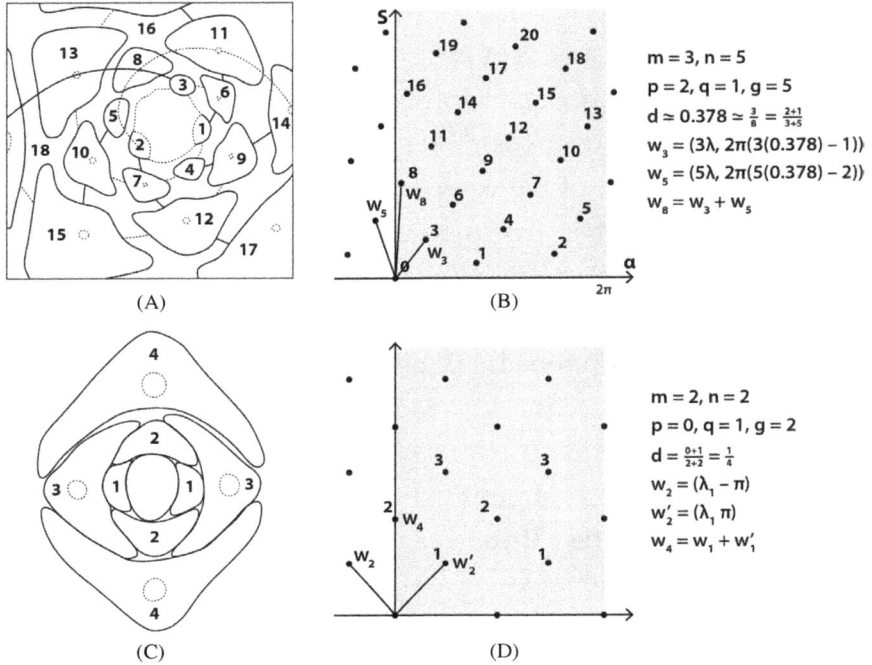

(A) (B)

$m = 3, n = 5$
$p = 2, q = 1, g = 5$
$d \simeq 0.378 \simeq \frac{3}{8} = \frac{2+1}{3+5}$
$w_3 = (3\lambda, 2\pi(3(0.378) - 1))$
$w_5 = (5\lambda, 2\pi(5(0.378) - 2))$
$w_8 = w_3 + w_5$

(C) (D)

$m = 2, n = 2$
$p = 0, q = 1, g = 2$
$d = \frac{0+1}{2+2} = \frac{1}{4}$
$w_2 = (\lambda, -\pi)$
$w_2' = (\lambda, \pi)$
$w_4 = w_1 + w_1'$

Fig. 5.4. Outline of the leaf bases of (**A**) *Sempervivum tectorum* and (**C**) *Honkenia peploides* numbered according to their order of appearance from younger to older (adapted from Church, 1904 and Rutishauser, 1998, respectively). (**B, D**) Representation of (**A**) and (**C**), respectively, in radial s and angular α coordinates. The points representing leaves form lattices. (Modified from Shipman and Newell, 2005; reprinted with permission from Elsevier.)

phyllotactic pattern represented in the lattice is described geometrically[1] by the vectors $\vec{w}_m$ and $\vec{w}_n$, where $m = 3$ and

---

[1] The generators of the phyllotactic lattice in Fig. 5.3B are given by the following equations:

$$\vec{w}_m = \frac{1}{g}(\lambda m, (md - q)) \text{ and }$$

$$\vec{w}_n = \frac{1}{g}(\lambda n, (nd - p)),$$

where d is the divergence angle, λ is the rise, m and n are a pair of parastichies (integers), and p and q are integers such that $pm - qn = \pm g$.

(A)

(B)

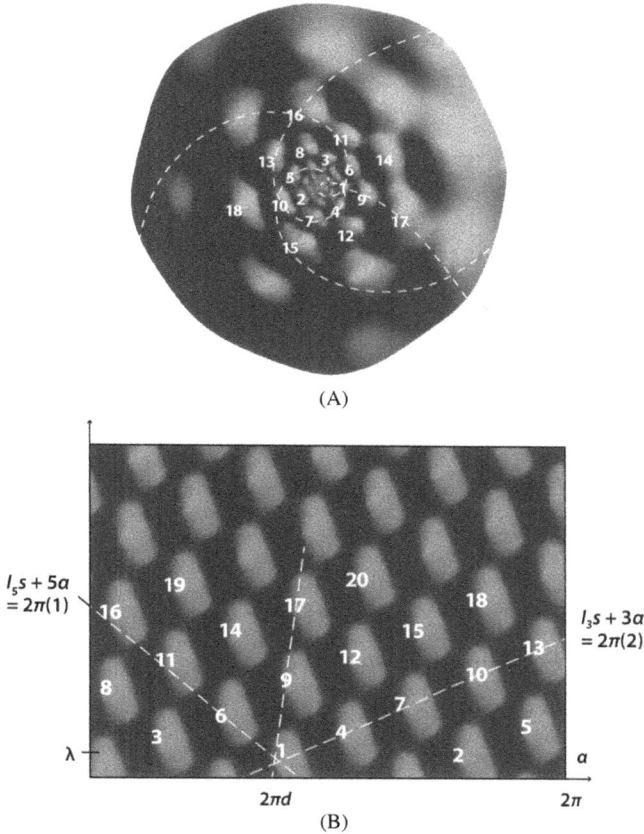

Fig. 5.5. **(A)** A deformation of the sphere given by the function $w(r, \alpha) = a_3 \cos(l_3 \ln(r) + 3\alpha) + a_5 \cos(l_5 \ln(r) + 5\alpha) + a_8 \cos(l_8 \ln(r) + 8\alpha)$, where $l_3 + l_5 = l_8 \approx 0$. **(B)** The same deformation plotted in the $(s, \alpha$-plane, where $s = r$. (From Shipman and Newell, 2005; reprinted with permission from Elsevier.)

$n = 5$. The same type of phyllotactic lattice (Fig. 5.5A,B)[2] corresponds to the vectors $\vec{k}_m$ and $\vec{k}_n$. The vectorial representa-

---

[2] In Shipman and Newell's model (2005), the wave vectors that generate phyllotactic patterns (Fig. 5.4A,B) are

$$\vec{k}_m = (l_m, m) = \left(\frac{2\pi}{\lambda}(q - md), m \right) \quad \text{and}$$

$$\vec{k}_n = (l_n, n) = \left(\frac{2\pi}{\lambda}((p - nd), n) \right).$$

tion (Fig. 5.5B) in contrast to the geometrical lattice (Fig. 5.4B) considers the shape of the primordia.

Equation Minimising Elastic Energy

Shipman and Newell (2005) and Newell and Shipman (2005) showed how the interactions between different wavevectors lead to observed phyllotactic patterns. Phyllotactic patterns result from (1) the strain energy, which is the product of the perturbed Airy stress function[3] (function that can be used to determine the stress on a structure) and Gaussian curvature[4] of the deformed surface, and (2) the presence of special triads of modes (wavevectors satisfying the condition $\vec{k}_1 + \vec{k}_2 = \vec{k}_3$), which minimise this energy. The authors defined 'special triads' as preferred wavevectors that maximise a combination of their linear growth rates $\sigma(\vec{k}_j)$ and an interaction coefficient $\tau(\vec{k}_1, \vec{k}_2, \vec{k}_3 = \vec{k}_1 + \vec{k}_2)$. The coefficient σ depends on the value of $\Gamma = \frac{R}{\Lambda}$, where R is the mean radius R of the generative region and Λ, the natural wavelength. Both σ and τ are related to specific physical parameters described in the section Parameters.

Newell and Shipman's analysis (2005) was aimed at determining the elastic-energy-minimising deformation (w) of the tunica in the generative region that generates phyllotactic patterns. They produced an equation representing the elastic energy $\mathfrak{E}(w, f)$, which is a function of the normal deflection of the tunica shell (w) and the Airy stress function (f) in the radial (r) and angular (α) directions.

---

[3] Function that can be used to determine the stress on a structure.

[4] In differential geometry, the Gaussian curvature is a measure of the curvature of a surface at a point. For example, the Gaussian curvature of a sphere is positive, that of a hyperboloid is negative, and that of a cylinder is zero.

$$\mathfrak{E} = (w, f) = \int \left[\begin{array}{c} \frac{1}{2}(\Delta w)^2 + V(w) - \frac{1}{2}P\left(\chi \partial_r w + \frac{1}{\Gamma^2}\partial_\alpha w \right)^2 \\ + f\left(C\Delta w - \frac{1}{2v\Gamma^2}[w,w] \right) - \frac{1}{2(\Delta f)^2} \end{array} \right] dr\, d\alpha, \quad (2)$$

where $[f, w] = f_{rr}w_{\alpha\alpha} + f_{\alpha\alpha}w_{rr} - 2f_{r\alpha}w_{r\alpha}$. Newell and Shipman (2005) assumed that the stresses are constant through the width h. If σ_{ij} denotes the stress tensor, then $f_{rr} = h\sigma_{\alpha\alpha}, f_{r\alpha} = -h\sigma_{r\alpha}$, and $f_{\sigma\sigma} = h\sigma_{rr}$.

In Eq. (2), the first term represents a resistance to buckling and corresponds to the bending energy of the shell. The second term, $V(w)$, is a potential coming from the elastic foundation and a pressure exerted by the growth of the corpus; it is equal to $\frac{\kappa}{2}w^2 + \frac{\gamma}{4}w^4$, where κ and γ are spring constants. The next two terms correspond to the strain energy, which is equal to the Airy stress function multiplied by the Gaussian curvature. The last term corresponds to in-surface deformations. The parameters involved in this equation are described in greater detail below.

In the solution for Eq. (2), the perturbation energy is related to the amplitude (A) of the wavevectors in the following equation[5] (Newell and Shipman, 2005)

$$\mathfrak{E} = -\sum_{\vec{k}\in\mathfrak{A}}\sigma_j\left(l_j, m_j\right)A_jA_j^* - \sum\tau_{123}\left(A_1A_2A_3 + A_1^*A_2^*A_3^*\right)$$
$$+ \sum_{c,d=1}^{N}\gamma_{cd}A_cA_c^*A_dA_d^*. \quad (3)$$

This equation is interesting because it shows how the physical parameters are linked to the wavevectors that generate

[5] Details of calculations are given in Shipman and Newell (2005).

phyllotactic patterns. In Eq. (3), the first sum comprises all vectors belonging to the active set $\mathfrak{U}$, which is the set of $\vec{k}_j(l_j, m_j)$ for which the value of the linear growth rates σ, which depend on (l_j, m_j), are above a given threshold. In other words, the active set corresponds to vectors that can be involved in the formation of phyllotactic patterns.[6]

The cubic terms correspond to all wavevector triads in the set $\mathfrak{U}$ where, for all triplets $\vec{k}_1, \vec{k}_2, \vec{k}_3$, we have $\vec{k}_1 + \vec{k}_2 = \vec{k}_3$. The coefficient $\tau_{123}(\vec{k}_1, \vec{k}_2, \vec{k}_3 = \vec{k}_1 + \vec{k}_2)$ is a function of these three vectors. The last terms (quartic) are positive and refer to the underlying and 'elastic' corpus. In this model, the system evolves to a minimum energy over time. Then, the goal is to find phyllotactic patterns minimising the energy $\mathfrak{E}$ of the system. To relate this theoretical model to naturally occurring patterns, Shipman and Newell (2005) showed that the particular coefficients $\sigma(\vec{k}_j)$, and $\tau(\vec{k}_r, \vec{k}_s, \vec{k}_{r+s} = \vec{k}_r + \vec{k}_s)$ have properties which lead to a range of phyllotactic patterns (phyllotactic lattices) and transitions between phyllotactic patterns and considered the shapes of appendages. By using this approach, they showed that the elastic energy of the system tends to be minimal when the dominant circumferential wave numbers (m, n) belong to the Fibonacci series.

Parameters

The many parameters appearing in Shipman and Newell's equations (2005) describe the material properties of the tunica (skin)

---

[6] More formally, Shipman and Newell (2005, p. 181) '...define the active set $\mathfrak{U}$ to be the set of all modes $\vec{k} = (l, m)$ such that $(l, m) > -3\sigma(l_c, m_c)$, where $\vec{k}_c = (l_c, m_c)$ is the (a) wavevector of the mode(s) with the largest positive linear growth rate. The choice of 3 is not important to the outcome; the factor must be large enough to include all relevant modes'.

and corpus (foundation), stresses in the tunica, and geometry of the pre-buckling shell. Although the values of certain parameters can be obtained experimentally (Dumais and Steele, 2000), the exact estimation of many of these parameters, such as the material properties of the tunica and corpus, remains problematic. However, the mathematical description of these parameters provides a good theoretical framework for future experimental studies.

In Eq. (2), some nondimensional parameters are defined by the natural wavelength $2\pi\Lambda$ of the shell, where

$$\Lambda^4 = \frac{Eh^3 v^2}{K + \frac{Eh}{R_0^2}}.$$

In this equation, E is Young's modulus[7]; h, the thickness of the tunica; and $v^2 = \frac{1}{12(1-\mu^2)}$, where u is the Poisson ratio of the shell. κ is a linear spring constant measuring the strength of the corpus foundation. However, no experimental value for κ in plants has been reported. For the other material parameters (E, h, u), Shipman and Newell (2005) used empirical values published by different authors (Hejnowicz and Sievers, 1996; Dumais and Steele, 2000; Steele, 2000).

Shipman and Newell (2005) reviewed the nondimensional parameters in the energy equations in detail. We do not aim to discuss the properties of these parameters in detail; however, a short description of these parameters could provide a better understanding of the tangible physical and geometrical parameters involved in the model in question.

---

[7]Young's modulus is a mechanical property of elastic solid materials. It is a measure of the relationship between stress (force per unit area) and strain (proportional deformation) in a material. It is expressed in units of pascals.

Corpus (tissue located below the tunica) foundation parameters:

$$\kappa' = \frac{\Lambda^4 \kappa}{Eh^3 v^2} \quad \text{and} \quad \gamma' = \frac{\Lambda^4 \gamma}{Ehv^2}.$$

A positive value of K (strength of corpus foundation) and γ (hard spring response to corpus foundation) implies that the corpus foundation plays an active role in the appearance of phyllotactic patterns. However, Shipman and Newell (2005) noted that we do not have a good estimate of these parameters.

Stress parameters

Dumais and Steele's experimental work (2000) on sunflower capitula suggests the presence of compressive stresses in the circumferential direction. Based on this study, Shipman and Newell (2005) deduced that the principal stresses (averaged through the width of the shell) are thus the radial stress $N_{rr} = h\sigma_{rr}$ and the circumferential stress $N_{\alpha\alpha} = h\sigma_{\alpha\alpha}$, where h is the width of the tunica shell and σ_{ij}, the tensor of in-plane stresses.

The corresponding dimensionless parameters are the compressive stress

$$P = -\frac{N_{\alpha\alpha}\Lambda^2}{Eh^3 v^2},$$

and the ratio of the radial stress on the circumferential stress

$$\chi = \frac{N_{rr}}{N_{\alpha\alpha}}.$$

The stress parameter P is the critical value at which unstable modes appear, resulting in conditions suitable for primordium initiation. Unfortunately, no experimental value is available for this parameter. In their model, Shipman and Newell (2005) assumed that N_a is negative; therefore, $P > 0$. Further, N_r is either positive or negative with a magnitude less than that of N_a; therefore, $\chi < 1$.

Geometric parameters

$$C = \frac{\Lambda^2}{R_a h \upsilon} \quad \text{and} \quad \rho = \frac{R_a}{R_r}.$$

These two parameters describe the curvature of the SAM where primordium initiation occurs. When $\rho = 1$ and $C > 0$, a spherical apex is seen (Fig. 5.3A); when $\rho = 1$ and $C < 0$, an inverted sphere is seen (Fig. 5.3B); and when $C > 0$ and $\rho < 0$, a hyperbolic geometry is seen (Fig. 5.3C). The parameter C (tunica curvature) is analogous to the conicity parameter in Douady and Couder's model (1996a). For large C values, the triads are ridge-dominated and a new ridge can be added every time Γ increases by an integer. In other cases, depending on the curvature C and the ratio of radial to circumferential stress χ, an approximately hexagonal platform will appear (Shipman and Newell, 2005).

The geometric dimensionless parameter $\Gamma = \frac{R}{\Lambda}$ is the ratio of the circumference $2\pi R$ of the central circle in the generative region to the natural wavelength $2\pi\Lambda$; this is roughly the inverse of Douady and Couder's parameter $\tilde{\Gamma}$, and it plays an analogous role in Shipman and Newell's model (2005). Γ increases with the SAM size (as R increases), and the biological and physical properties of the SAM remain constant as is the case for the wavelength (Λ). Variations in Γ lead to the appearance of different patterns (Table 5.1). If buckling occurs at radius R and wavelength Λ, Γ should peak in a spatially periodic ripple. Shipman (2010) analysed the mathematical

Table 5.1. Energy-minimising parastichy numbers for small Γ values

Γ	m	n	$m+n$	Type of pattern
2	1	1	2	Alternating 1-whorl
3	1	2	3	Fibonacci spirals
4	2	2	4	Alternating 2-whorl
5	2	3	5	Fibonacci spirals
6	3	3	6	Alternating 3-whorl
7	3	4	7	Lucas spirals
8	4	4	8	Alternating 4-whorl
9	4	5	9	
10	5	5	10	Alternating 5-whorl
11	5	6	11	
12	6	6	12	Alternating 6-whorl

relationship between Γ and the primordium shape in phyllotactic organisation.

In the mathematical formulations of Shipman and Newell's model (2005), the constraint $C + \kappa' = 1$ reduces the number of parameters in the energy equations (Eq. (2)) to seven (γ, P, χ, C, ρ, Γ, and v). When $\rho = 1$, the number of parameters is reduced to four.

Area of Newly Formed Primordia

Shipman and Newell's model considers the area and form of newly formed primordia, and they developed a formula to calculate the area of a primordium at its initiation. To approximate the area, they used the vectors $w_1' = \frac{1}{g}(\lambda m, 2\pi R(md - q))$ and $w_2' = \frac{1}{g}(\lambda n, 2\pi R(nd - p))$ and scaled the radial coordinate by the wavelength Λ, which is one of the main variables involved in the generation of phyllotactic patterns. The area in question corresponds approximately to the parallelogram determined by the vectors

$$w_1'' = \frac{1}{g}(\lambda\Lambda m, 2\pi R(md - q))$$

and

$$w_2'' = \frac{1}{g}(\lambda\Lambda m, 2\pi R(nd - p)).$$

The area will then be equal to the absolute value of the determinant of the matrix,

$$\Omega'' = \frac{1}{g}\begin{pmatrix} \lambda\Lambda m & \lambda\Lambda n \\ 2\pi R(md - q) & 2\pi R(nd - p) \end{pmatrix},$$

which is $A = |\det \Omega''| = \frac{2\pi}{g}R\Lambda\lambda = \frac{2\pi}{g}\Gamma\Lambda^2\lambda$. This formula describes the primordium size at initiation. However, during the growth of the apex, primordia continue to grow and drift away from the centre of the apex. Based on this model, the primordium size increases exponentially if the radial growth of the SAM is exponential or at a constant rate if the radial growth of the SAM is constant.

Planform

Newell and Shipman (2005) simulated the formation of ridges and various polygonal configurations (hexagons and parallelograms) depending on the values of the parameters involved in the equations of their model. Shipman (2010) also analysed shape invariance in *phyllotactic planforms* in detail from a geometric viewpoint in phyllotactic cylindrical lattices and in the more general dynamic context of partial differential equations (PDE).

Further, outside the region denoted as R, no further change occurs in the form of deformation (i.e. organ) except that the amplitude may grow uniformly. This is different from a real plant in which the form of appendices can change during growth. Shipman and Newell (2005) used the term polygonal planform to designate the tiling of primordia on the surface of the stem or inflorescence axis as irregular polygons that usually correspond to ribs, parallelograms, hexagons, and so-called staircase parallelograms (Fig. 5.1).

The value of energy minimising the amplitude A_j is the theoretical calculation for the formation of a buckling pattern in the generative region (Newell and Shipman, 2005). Depending on the amplitude A_j, one can obtain ridged patterns (alternating 2-whorls) or spiral patterns with hexagonal or parallelogram patterns.

To explain the shape of primordia, they assigned a weight based on amplitude a_j to the lattice generators. To establish a relation between the shape of the primordia and phyllotactic patterns, they used the equation

$$w(s, \alpha) = \sum_{j=1}^{N} a_j \cos(\vec{k}_j \cdot \vec{x}), \qquad (4)$$

where wavevectors $\vec{k}_j = (l_j, m_j), \vec{x} = (s, \alpha)$, and $\vec{k}_j \cdot \vec{x} = l_j s + m_j \alpha$. Various combinations of amplitudes a_j and wavevectors $\vec{k}_j$ were used to generate plots from the equation.

The phyllotactic patterns and shape of primordia in Fig. 5.5 correspond to different combinations of wavevectors and amplitudes. Shipman and Newell (2005) stated the following (p. 160):

(i) patterns with dominant ridges are produced by a sum $w(s, \alpha) = \sum_{j=1}^{3} a_j \cos(\vec{k}_j \cdot \vec{x})$ in which $\vec{k}_1 + \vec{k}_2 = \vec{k}_3$ and $a_3 > a_1 \simeq a_2$ (Fig. 5.1E),

(ii) hexagonal planforms are produced by a sum $w(s, \alpha) = \sum_{j=1}^{3} a_j \cos(\vec{k}_j \cdot \vec{x})$ in which $\vec{k}_1 + \vec{k}_2 = \vec{k}_3$ and all amplitudes a_j are approximately equal (Fig. 5.1F),

(iii) parallelogram planforms are produced by a sum $w(s, \alpha) = \Sigma_{j=1}^{4} a_j \cos(\vec{k}_j \cdot \vec{x})$ in which $\vec{k}_1 + \vec{k}_2 = \vec{k}_3$, $\vec{k}_2 + \vec{k}_3 = \vec{k}_4$ and $a_1 \simeq a_4 < a_2 \simeq a_3$ (Fig. 5.1G),

(iv) staircase parallelogram planforms are produced by a sum $w(s, \alpha) = \Sigma_{j=1}^{5} a_j \cos(\vec{k}_j \cdot \vec{x})$ in which $\vec{k}_1 + \vec{k}_2 = \vec{k}_3$, $\vec{k}_2 + \vec{k}_3 = \vec{k}_4$, $\vec{k}_3 + \vec{k}_4 = \vec{k}_5$, and $a_1 \simeq a_5 < a_2 \simeq a_4 < a_5$ (Fig. 5.1H).

Transitions

During plant growth, continuous and discontinuous transitions can occur between phyllotactic patterns (see Chapter 1). In their model, Shipman and Newell (2005) recognised discontinuous (first order) and continuous (second order) transitions. There are two types of discontinuous transitions: (I, 1) where only one integer is shared by two triads, for example, $(2, \mathbf{3}, 5) \rightarrow (3, \mathbf{3}, 6)$, and (I, 2), which is rarer, where the two triads share two integers, for example, $(\mathbf{2}, 3, \mathbf{5}) \rightarrow (\mathbf{2}, 4, \mathbf{5})$. In continuous transitions, the values of some parameters such as Γ change continuously. Theoretically, two types of transitions exist. However, the type in which only one integer is shared by two triads (II, 1) has not yet been observed or described in nature. The other type, in which two triads share two integers (II, 2), for example, $(2, \mathbf{3}, \mathbf{5}) \rightarrow (3, \mathbf{5}, 8)$, is much more common.

Transitions of type (I, 1) are observed particularly in plants characterised by the presence of ridges (i.e. *Saguaro cacti*). They can also occur in plants with hexagonal configurations, such as pine cones or pineapple. This type of transition is also frequent in elongated inflorescences of Araceae, where there are discontinuous transitions between the number of parastichies from the basal part to the apical part (Jean and Barabé, 2001). Transitions of type (I, 1): $(2, 2, 4) \rightarrow (2, 3, 5)$ are explained by three wavevectors:

$$\vec{k}_m, \vec{k}_n, \vec{k}_{m+n} = \vec{k}_m + \vec{k}_n$$

Transitions of type (II, 2): (2, 3, 5) → (3, 5, 8) are usually observed at higher Fibonacci numbers, and the planform looks more like parallelograms than hexagons. Newell and Shipman (2005) noted that a continuous transition from one parallelogram planform to another is achieved through a continuous change in amplitude and divergence angle. They explained these transitions by invoking the presence of four or five wavevectors instead of three (single triad).

In their model, Shipman and Newell (2005) formalised physical principles acting at the SAM level that were characterised by biologists like Green based on experimental and observational studies. One of the novel aspects of their model is that primordia of different sizes and forms are produced depending on the natural wavelength of the phyllotactic pattern. This model also enables analysing quantitative interrelationships between the types of planforms and phyllotaxis. However, the model has many parameters whose empirical value is very difficult to obtain. Further, these parameters may be different for different plant species. Overall, the model plausibly reproduces different phyllotactic patterns in relation to the form of primordia. However, as with any other model, does the correspondence between the model and real patterns and form imply that this model reveals the biological processes involved in plant morphogenesis? Irrespective of whether a relationship exists, it is often difficult to elucidate biological processes in living systems (see concluding remarks in Chapter 6). However, the model provides a theoretical framework to empirically analyse the value of physical parameters involved in SAM formation. The next section discusses how Newell *et al.* (2008a,b) showed how the biomechanical model can be coupled with a biochemical model by using similar types of equations in both cases.

Coupling Mechanical and Biochemical Processes

In the past decade, studies have actively explored the relationship between physical and biochemical processes acting at the cellular level in relation to pattern formation (Landrein and Hamant, 2013; Hamant, 2017). Hamant *et al.* (2008) proposed that morphogenesis at the SAM level is controlled by two main regulatory circuits: biomechanical and biochemical. The first one corresponds to cell reinforcement against maximal tension directions along the meristem surface by a cell-autonomous mechanism governing the reorientation of microtubules. The second one corresponds to the synthesis and polar transport of auxin resulting in different concentration peaks on the SAM surface. They proposed that these two types of processes would operate partially independently or in parallel. Below, we present two models integrating physical and biochemical processes. The first incorporates auxin transport by PIN1 and mechanical buckling to generate the appearance of phyllotactic patterns. The second one shows how the interaction between physical stress and auxin concentration at the cellular level can create particular patterns.

In dynamic models based on PDE, an initially homogenous state is broken as a control parameter crosses a threshold value. This broken symmetry gives rise to different organised patterns determined by universal symmetries as well as the parameters of the mechanisms involved (Shipman, 2010). Newell *et al.* (2008a,b) highlighted the theoretical and geometrical relationships underlying two dynamical models used in phyllotaxis: a biomechanical model and a biochemical model. The equations proposed for modelling both auxin transport by PIN1 proteins (e.g. Kuhlemeier, 2007; Nakayama *et al.*, 2012; see Chapter 4) and mechanical buckling models are very similar in form but differ in the interpretation of parameters (Newell *et al.*, 2008a). The equations for these two models are based on a critical value represented by either auxin-transport parameter H or compressive stress P. The authors demonstrated that these two models can be reduced to similar sets of differential equations of periodic functions of the form $wt + \Delta^2 w + P\Delta w + \Delta^4 w +$ nonlinear terms $= 0$, where w represents the normal deflexion of the buckling plant tunica (biophysical model) or auxin concentration (biochemical model) and P and Λ^4 are constants. Given that quadratic nonlinear terms are present in both models, they differ only by the interpretation of linear terms. P is the control parameter corresponding to the compressive stress in the biophysical model and the relative strength of auxin transport and diffusion in the biochemical model (Shipman, 2010). In the mathematical model of Newell *et al.* (2008a), the biochemical (auxin concentration) and biophysical (stress distribution) mechanisms can interact positively or negatively. Depending on the choice of parameters, the two mechanisms may act equally in the determination of patterns or one or the other mechanism may dominate.

Additionally, apex growth produces compressive stress in an annular generative region at the SAM periphery where primordia are initiated. As the tunica in the generative region cannot expand in the circumferential direction, this causes surface buckling in this zone. However, a nonuniform auxin concentration can also lead to corresponding changes in local stresses. The biophysical mechanism (stress) and the biochemical mechanism (auxin concentration) work in synergy. For example, when the PIN1 transport coefficient and the circumferential stress that is induced by growth in the generative region are what the authors describe as 'near-critical', the cooperation between these two components is believed to be very strong.

To mathematically represent the relationships between the local auxin concentrations and surface deformation, Newell *et al.*

(2008a,b) used continuous field variables. They derived a continuous approximation of the discrete biochemical model of Jönsson *et al.* (2006) for primordium formation. By using this reformulation in a continuous form, they analysed combined effects of auxin concentration and stress on the emergence of phyllotactic patterns. By combining the biochemical model of Jönsson *et al.* (2006) with their buckling model, Newell *et al.* (2008a) obtained the following results:

(i) When there is only one mechanism (biochemical or mechanical) that creates instability, it governs the choice of phyllotactic patterns; this represents a passive coupling of biochemistry and mechanics.

(ii) The coupling between biochemistry and mechanics is active when both mechanisms reach their instability thresholds, resulting in ridge-like surface deformations.

(iii) The formation of polygons of various shapes with a Fibonacci spiral phyllotactic lattice occurs when the coupling is active via nonlinear resonance (i.e. it depends on the amplitude of oscillations).

Newell *et al.* (2008a) reformulated the model of Jönsson *et al.* (2006), resulting in the following PDE for the auxin concentration field $g(x, y, t)$:

$$g_t + D_g \nabla^4 g + H \nabla^2 g + dg + \kappa_1 \nabla(g \nabla g) + \kappa_2 \nabla(\nabla g \nabla^2 g) = 0, \quad (5)$$

with

$$D_g = \frac{PTA_0^2 H_0^4}{4(\kappa + A_0)^2}, \quad H = -h_0^2 D_g + \frac{h_0^2 PTA_0^2}{(\kappa + A_0)^2},$$

$$\kappa_1 = \frac{2PTh_0^2 \kappa A_0}{(\kappa + A_0)^3} f_0, \quad \kappa_2 = \frac{PTh_0^4 \kappa A_0^2}{2(\kappa + A_0)^3} f_0.$$

A_i is the auxin concentration in the ith cell; A_0, the mean auxin level in the cells; f_0, the coefficient of cubic expansion; D_g, the diffusion coefficient; $g(x, y, t)$, the PDE for the auxin fluctuation field; h_0, the cell length in a one-dimensional lattice; H, the effective diffusion between cells; P, the total amount of PIN1 in a single cell that is assumed to be constant throughout all cells; $\kappa = \frac{k_2}{k_1}$, as listed in the equations of Jönsson *et al.* (2006); and T, a constant corresponding to the strength of dependency of auxin transportation on PIN1 distribution.

The authors showed how the patterns for both auxin concentration fluctuation and surface deformation fields interact if some coupling is included between the biochemical and the mechanical processes. For this purpose, Newell *et al.* (2008a) used the mechanical model involving stress–strain relations in the generative region (see preceding section). They denoted in-plane stresses (average width of tunica) by N_{ij} and total strains by ε_{ij}. Then, the stress–strain interactions in the angular (α) and radial (r) planes are respectively

$$N_{\alpha\alpha} = Eh(\varepsilon_{\alpha\alpha} - G_{\alpha\alpha}) \quad \text{and} \quad N_{rr} = Eh(\varepsilon_{rr} - G_{rr}),$$

where E is Young's modulus of the tunica and h, its thickness. Considering the stresses N_{ij} and the normal deflection $w(r, \alpha)$ of the tunica shell, the stress-equilibrium equation is

$$D\nabla^4 w - N_{rr}\frac{\partial^2}{\partial r^2}w - N_{\alpha\alpha}\frac{1}{R^2}\frac{\partial^2}{\partial \alpha^2}w - 2N_{r\alpha}\frac{1}{R}\frac{\partial^2}{\partial r\partial \alpha}w = 0, \quad (6)$$

In this equation, $D = Eh^3v^2$ and the bending modulus $v^2 = \frac{1}{12(1-u^2)}$ for Poisson's ratio u.

Newell *et al.* (2008a) used Eqs. (5) and (6) to produce a combined model consisting of the stress equilibrium equation (with an extra term that is proportional to the curvature of the spherical apex which includes the generative region) as a compatibility equation that relates f (an Airy stress function), g (auxin concentration field), and w (tunica deflection) to the biomechanical equations:

$$\zeta_m w_t + \nabla^4 w + P\left(\frac{1}{\Gamma^2}\frac{\partial^2}{\partial\alpha^2}w + \frac{\partial^2}{\partial r^2}w\right) + C\nabla^2 f - [f, w]$$

$$+ \kappa w + \gamma w^3 = 0 \tag{7a}$$

$$\nabla^4 f + \nabla^2 g - C\nabla^2 w + \frac{1}{2}[w, w] = 0 \tag{7b}$$

$$\zeta_{ij} g_t + \nabla^4 g + H\nabla^2 g + g + \nabla(\nabla g(\kappa_1 g + \kappa_2 \nabla^2 g)) - \beta\nabla^2 f + \delta g^3 = 0 \tag{7c}$$

In these equations, ζ_m and ζ_{ij} represent the 'relaxation time scale' of the elastic surface of the SAM and auxin field, respectively. In $\Gamma = \frac{R}{\Lambda_g}$, $2\pi\Lambda$ corresponds to the natural instability wavelength for the critical value of the auxin flux H (i.e. $H_c = 2$) when $\Lambda_g^4 = \frac{D_g}{d}$. C is the nondimensional curvature corresponding to $\frac{\Lambda_g^2}{R_c v h}$, where R_c is the nondimensional curvature of the spherical SAM, $v = \sqrt{\frac{1}{12(1-\mu^2)}}$, and h is the tunica thickness. u is the Poisson ratio of the shell. The interaction between the tunica and the underlying corpus is characterised by nondimensional spring constants (related to the material stiffness) κ and γ. These constants are related to their dimensional forms K' and γ' by $K = \frac{\Lambda_g^4}{D}K'$ and $\gamma = \frac{\Lambda_g^4 v^2 h^2}{D}\gamma'$. The natural wavelength for mechanical buckling is $\Lambda_m^4 = \frac{DR_c^2 v^2 h^2}{D + \kappa' R_c^2 v^2 h^2}$ and, as outlined by Newell *et al.* (2008a), if $L = \frac{\Lambda_g^4}{\Lambda_m^2}$, the constraint $\kappa + C^2 = L$ is satisfied. $[w, w]$ is closely related to the Gaussian curvature of the deformed elastic surface. Based on Dumais and Steele's experimental data (2000), Newell *et al.* (2008a) determined that the in-plane stress only acts in the α direction; consequently, the ratio of the radial stress (N_{rr}) to the circumferential stress ($N_{\alpha\alpha}$) χ is zero.

In summary, Newell *et al.* (2008a,b) showed that both biomechanical (surface deformation) and biochemical (auxin concentration) mechanisms interact to produce phyllotactic patterns. They recognised four cases:

Case 1. Auxin mechanism dominates. In this case, surface deformation will look very similar to auxin deformation profiles. Spiral patterns are initiated.

Case 2. Mechanical mechanism dominates. This is the opposite of case 1. Here, the resulting whorl pattern governs the fluctuation in auxin concentration. Whorled patterns are initiated.

In cases 1 and 2, the amplitudes for the other mechanisms are governed and match those of the dominant mechanism.

Case 3. Cooperation between the two mechanisms via linear resonance. In this case, both auxin concentration and compressive stress have common maxima and are closer in terms of their critical values. Whorled patterns are initiated.

Case 4. Cooperation between the two mechanisms via nonlinear resonance can also occur through the nonlinear terms in Eqs. (7a) and (7b). These nonlinear interactions result in Fibonacci spiral patterns.

Newell *et al.* (2008a) ultimately predicted not only the structure of the phyllotactic lattice but also the corresponding tiling planform. The parameter $L = \frac{\Lambda_g^4}{\Lambda_m^4}$, the ratio of biochemical to mechanical natural wavelength determines the relation between the phyllotactic lattice and the tilling planform. For example, if $L = 5$, the biochemical wavelength is greater than the biophysical one ($\Lambda_g = 1.5\Lambda_m$) and we obtain a spiral pattern with elongated diamond-shaped primordia (Fig. 5.1D). With regard to the shape of primordia, Newell *et al.* (2008a) noted that multiple mechanisms can produce the difference between the two types of parastichies: (1) contact, connecting touching primordia, and (2) conspicuous, connecting nearest dots denoting the position of primordia (Shipman, personal communication). Strickland *et al.* (2016) recently analysed this problem; they used the preceding model to mathematically analyse the formation of square lattices in pattern-forming systems such as those observed in Cactaceae.

Rueda-Contreras *et al.* (2018) developed another type of model combining physical and biochemical principles. They developed a mathematical model in which they assume that the auxin concentration modifies the mechanical properties of the SAM; in turn, the mechanical stress field experienced by the tissue governs the direction of auxin flux. In this model, a chemical reaction-diffusion process is coupled with the curvature and the mechanical stress of a three-dimensional domain in a feedback interaction. The auxin concentration promotes surface growth by enhancing the local curvature of the domain. This modification changes the geometry and mechanical properties (stress and strain) of the SAM. The relaxation of the surface rigidity will allow the emergence of a new primordium. Rueda-Contreras *et al.* (2018) used this model to reproduce different whorled patterns, ribbed patterns, and aberrant patterns. They theoretically showed how aberrant patterns can emerge from ribbed or whorled patterns by varying the value of two parameters: stress strength on the auxin flux and strength of the influence of auxin on the local curvature. However, they did not analyse the transition (continuous or discontinuous) between different spiral patterns that are known to occur frequently in nature.

Integration of Physical and Biochemical Parameters at the Cellular Level

Although Heisler *et al.* (2010) did not develop a model predicting the emergence of different phyllotactic patterns using physical or biochemical parameters, their modelling of the interaction between physical stress and auxin concentration at the cellular level provides a strong foundation that could be integrated into the theoretical study of phyllotaxis.

Morphogenesis is regulated by molecules like PIN1 involved in intercellular signalling and by the mechanical properties of cell walls. Experimental studies and microscopic observations suggest that mechanical signals play a role in regulating the orientation of microtubules and PIN1 concentration (Heisler *et al.*, 2010; Sassi *et al.*, 2014). Although Heisler *et al.* (2010) did not develop their own phyllotactic model, mathematically linking the physical stress experienced by the cell wall to the amount of PIN1 molecules in the cell constitutes a new approach to analyse a phyllotactic organisation. They hypothesised that each cell in the epidermal tissue of the SAM is subjected to tensional mechanical forces that influence PIN1 localisation. As a result of local auxin-induced growth, the expansion of a cell is perceived by its neighbour(s), resulting in PIN1 accumulation in the plasma membranes adjacent to the expanding cell. This ultimately leads the cell to export auxin to that expanding neighbour, thereby increasing its auxin concentration and contributing further to expansion and creating a feedback loop between cell wall strength and auxin concentration. Shi and Vernoux (2019) discussed the regulation of chemical and mechanical signals on phyllotaxis, which also includes the role of cytokinin, another important growth regulator involved in patterning at the SAM level.

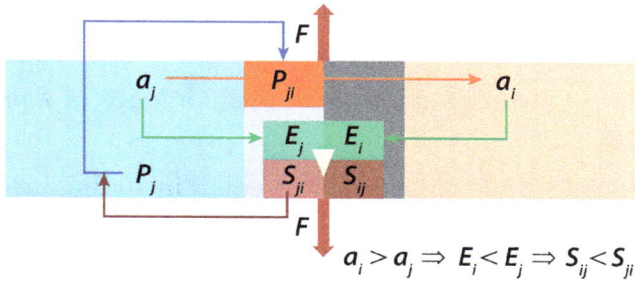

Fig. 5.6. Mathematical model of auxin transport and mechanical stress. Schematic representation of the interactions leading to a pattern-forming behaviour in the model. Auxin (a) is transported out of cell j to cell i by the PIN1 (P) proteins localised to the membrane, P_{ji}. Auxin concentration in each cell affects the elasticity (E_i and E_j) of the adjacent wall, which influences the mechanical stress ($S_{ij} < S_{ji}$) perceived in both parts of the wall between the cells as a result of the force F. For example, $a_i > a_j$ leads to $E_i < E_j$, in turn causing $S_{ij} < S_{ji}$. The cycling of PIN1 between cytosol, P_j, and the membrane, P_{ji}, depends on these stresses, and larger S_{ij} causes stronger allocation of P_j to P_{ji}. (From Heisler *et al.*, 2010.)

The dynamic of this model is similar to that of the model proposed by Jönsson *et al.* (2006) where PIN1 localisation is governed by the auxin concentration in neighbouring cells. However, in this new model, the dynamic of PIN1 proteins depends on mechanical stresses acting on the cell walls rather than the auxin concentration in neighbouring cells. The model of Heisler *et al.* (2010) is based on both mechanical and chemical interactions within single cells or between wall compartments of neighbouring cells (Fig. 5.6).

The mathematical model based on the one published by Hamant *et al.* (2008) combines mechanical and biochemical interactions in cells and cell wall compartments in the epidermal (L1) layer of the shoot tissue. Heisler *et al.* (2010) assumed that the tunica (L1) is under tension due to turgor pressure and interaction with inner tissues. The high level of stress in the cellular wall induces

PIN1 relocation to the adjoining membrane. A positive feedback loop exists between auxin concentration in the cell and auxin transport to this cell, which can lead to the emergence of a pattern of auxin peaks.

The change in auxin level a_i in a cell i, including production, degradation, passive transport, and PIN1-dependent active efflux of auxin, is given as

$$\frac{da_i}{dt} = c_a - d_a a_i + \sum_{k \in N_i} D(a_k - a_i) + \sum_{k \in N_i} (P_{ki} h(a_k) - P_{ik} h(a_i)) \qquad (8)$$

where c_a is the production constant; d_a, the degradation constant; and D, the passive transport constant. The formulation of Eq. (8) by Heisler *et al.* (2010) is based on the chemiosmotic transport theory of Sahlin *et al.* (2009), where the constants are linked to the permeability of the active transport term, such that $D \propto \frac{p_{aH}}{p_{pin}}$, where p_{ah} is the passive permeability of the protonated form of auxin and p_{pin}, the permeability of the PIN1-dependent efflux. The saturation point for auxin in active transport is described by a Michaelis-Menten function $h(a_i)$:

$$h(a_1) = \frac{a_i}{K_a + a_1} \qquad (9)$$

Heisler *et al.* (2010) assumed that PIN1 is in quasi-equilibrium. The function P_{ij} describes the PIN1 protein concentration resulting from stress. This stress causes the cycling of the PIN1 protein between the cytosol and the membrane in the wall separating cells i and j:

$$P_{ij} = \frac{P f(s_{ij})}{1 + \sum_{k \in N_i} f(s_{ik})}. \qquad (10)$$

P represents the total number of PIN1 molecules in the cell. The function $f(s_{ij}) = f_{exo}(s_{ij})/f_{endo}(s_{ij})$ corresponds to the response of auxin to stress. The authors describe it as a combination of increased exocytosis ($f_{exo}(s_{ij})$) and/or decreased endocytosis ($f_{endo}(s_{ij})$). They chose to use

$$f(s_{ij}) = k_2(s_{ij})^n, \tag{11}$$

where k_2 and n are constants and s_{ij} is the measured stress in the wall between cells i and j. The auxin-dependent elastic modulus of the wall of cell i is given as

$$E(a_i) = E_{min} + \frac{(E_{max} - E_{min})k_3^m}{a_i^m + k_3^m}, \tag{12}$$

where E_{min} and E_{max} are respectively the expected minimal and maximal values of wall elasticity, and k_3 and m are constant parameters (Hamant *et al.*, 2008). If cells are organised on a regular lattice, where the wall length is L_0, cross-sectional cell area is A_0, and isotropic force on each wall is F_0, then the strain of the composite is

$$e_{ij} = \frac{F/A_0}{E(a_i) + (a_j)}, \tag{13}$$

and the stress perceived on one side (s_{ij}) of the wall is

$$s_{ij} = \frac{F/A_0}{1 + \dfrac{E(a_j)}{E(a_i)}}. \tag{14}$$

By combining Eqs. (10) and (14), Heisler *et al.* (2010) showed a clear theoretical relationship between stress and auxin concentration (for more details, see their original paper).

The simulations performed by Heisler *et al.* (2010) indicate that a mechanism in which PIN1 localisation is governed by cell wall stress can generate phyllotactic-like patterns. Although the precise regulatory mechanism remains unclear, there is a tight coupling between PIN1 localisation and microtubule array orientation. This result will certainly open new avenues of research in regard to morphogenesis in general and phyllotaxis in particular.

Conclusion

The phyllotactic model of Shipman and Newell (2005) focuses on physical processes hypothesised to occur at the SAM level. Although a few experimental studies suggest that stress tensions act at the SAM level, there is no direct evidence that this physical element plays an active role. However, the mechanical model is novel in that it combines physical and geometrical variables that could theoretically act at the SAM level in the context of a global dynamic structure. The mechanical models, like the biochemical models (see Chapter 4), are based on various complex parameters linked to the SAM geometry and the interaction between developmental processes. In most cases, the empirical value of these parameters remains to be determined. Consequently, this model essentially serves as a theoretical framework for a detailed analysis and empirical measurement of many parameters controlling the dynamics of the system that generates patterns. Further, it should be noted that physical and biological processes acting at the cellular level play an active role in SAM development and the appearance of phyllotactic patterns. Thus, although it is possible to include mechanical and biomechanical processes in the same dynamical system, as shown by Newell *et al.* (2008a,b), the problem of linking the cellular level to the macroscopic level in the same biomechanical model remains unsolved.

6 Concluding Remarks: Critical Analysis and State of the Discipline

Chapter

As outlined in this book and other studies cited herein, many different types of phyllotactic patterns are seen across many different taxa. The central focus of phyllotactic studies has been to elucidate the mechanisms underlying the initiation and arrangement of lateral organs on meristems. However, the evolutionary context with regard to the universality of the manifestation of phyllotaxis has been less explored. Gola and Banasiak's recent review (2016) on the diversity and stability of phyllotactic patterns in major lineages of land plants provides a good cross-sectional comparative perspective. One overarching theme or observation is that similar types of phyllotactic patterns exist in phylogenetically distant groups of plants, suggesting a universal process in operation across the plant kingdom. Although the information on some lineages is incomplete, there are interesting trends worth noting. These trends are discussed in the context of the SAM structure, which varies from a single apical cell in bryophytes to layered multicelled meristems with specific zonation patterns in seed plants (Fig. 6.1).

Phyllotaxis: Evolutionary Considerations

Three phyllotactic patterns characterise bryophytes: spiral phyllotaxis as well as distichous and tristichous patterns, all of which are associated with a single apical cell and its potential to divide into various planes (Fig. 6.1C). Spiral and whorled phyllotactic patterns have also been reported in lycophytes, another group with a relatively basic SAM (one or two apical cells). In eusporangiate ferns, where the SAM is also represented by an apical cell and its derivatives or segments (merophytes), whorled, distichous, and spiral patterns have been documented. Gymnosperms and angiosperms are characterised by SAMs with distinctive zonation patterns, and consequently, a relatively larger number of initial cells (Figs. 6.1A–B). Most phyllotactic patterns in those groups are also represented by spiral (predominantly Fibonacci) and whorled arrangements (including distichy in angiosperms). However, variations in the relationship between the size of the peripheral (organogenetic) zone and the size of the primordia in seed plants create conditions for a relatively broader range in the diversity of patterns observed. Changes in leaf arrangement in the context of a single plant are common in many lineages and are tied to changes in the shape of the apical cell or SAM geometry.

However, according to Gola and Banasiak (2016), the SAM geometry is likely only one of several factors implicated in the manifestation of phyllotactic patterns. The well-explored role of auxin in phyllotaxis regulation (e.g. Smith *et al.*, 2006b) is establishing its universality as a mechanism in seed plants. However, another aspect of pattern formation is also believed to play a role in the determination of phyllotactic patterns: the relationship between the flexibility of a phyllotactic pattern and the stability of the internal (stellar) vascular system, as noted by Gola and Banasiak (2016). Banasiak (2011) noted the complementarity between the vascular

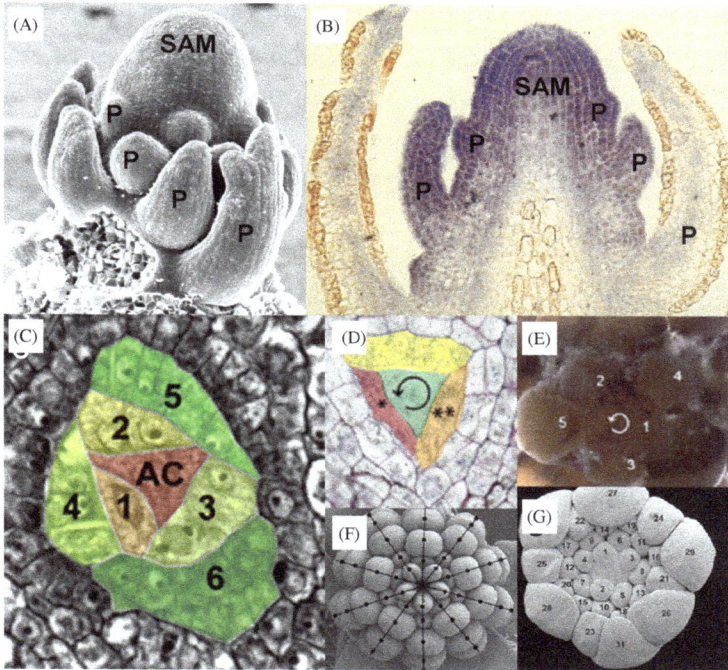

Fig. 6.1. Examples of phyllotactic patterns across taxa. **(A)** Scanning electron microscope (SEM) photo of a side view of SAM of *Myriophyllum aquaticum* showing leaf primordia (P) at various stages of development (40×). **(B)** Longitudinal section of the shoot tip of *M. aquaticum* showing leaf primordia (P) and the 'tunica-corpus' pattern of cell layering of the SAM typical of angiosperms (40×). **(C)** Transverse section of the SAM of bryophyte *Atrichum undulatum* showing the apical cell (AC) and 'successively cut-off segments' referred to as merophytes (numbered from most recently formed (1) to successively older segments (2–5). (From Gola and Banasiak, 2016; no magnification provided.) **(D)** Transverse section of the growing tip of *Sargassum muticum* showing the counterclockwise orientation of apical cell division and the location of the youngest (*) and the next youngest (**) daughter cell. (From Linardic and Braybrook, 2017; 320×.) **(E)** Top view of the growing tip of *S. muticum* with a counterclockwise spiral phyllotactic pattern showing sequence of primordia from youngest (1) to oldest (5). (From Linardic and Braybrook, 2017; 24×.) **(F)** SEM photo of top view of the flower of *Aquilegia vulgaris* L. (Ranunculaceae) with a whorled phyllotactic pattern. (From Endress and Doyle, 2007; no magnification provided; reprinted with permission from Elsevier.) **(G)** SEM photo of top view of the flower of *Austrobaileya scandens* CT white (Austrobaileyaceae) showing a spiral phyllotactic pattern with primordia numbered from youngest (1) to oldest (31). (From Endress and Doyle, 2007; no magnification provided.)

and L1 layer pathways of auxin transport and their involvement in organogenesis. He only observed complete inhibition of organogenesis in *Arabidopsis* when both pathways were impaired. Aulenback's study (2013) on stellar evolution and morphology, called vascullotaxy, may clarify the relationships between the establishment of the vascular system and the manifestation of phyllotactic patterns from an evolutionary perspective. The basic premise of this approach focuses on the fact that the 'axial sympodial arrangement' of vasculature should be used to interpret phyllotaxy and not the other way around.

As outlined throughout this book, the role of PIN proteins as auxin efflux facilitators is relatively well established, as demonstrated in recent empirical studies and models (Chapter 4). From an evolutionary perspective, Paponov *et al.* (2005) suggested that PIN genes, represented by a growing family of auxin efflux facilitators in higher plants, diverged from a single ancestral sequence. The confirmed presence and ubiquity of these genes throughout the plant kingdom, including in nonvascular plants, supports their close functional relationships and complementation in the context of basic developmental processes such as organogenesis. Nonetheless, within specific groups of plants such as, for example, monocots, the PIN family of proteins has a more divergent phylogenetic structure in comparison to the 'broadly' conserved structure in dicots (Paponov *et al.*, 2005). These differences in sequences may be related to specific morphogenetic processes involved in the generation of phyllotactic patterns that typically characterise monocots.

Despite obvious differences in development and types of organs formed in relation to the SAM, floral systems show a variety of phyllotactic patterns represented mainly by spiral and whorled arrangements across taxa (Fig. 6.1F–G). Recent phylogenetic analyses revealed the lability of floral phyllotactic patterns early during the evolution of angiosperms and the fact that spiral patterns in Magnoliales and Laurales may be derived rather than primitive (Endress and Doyle, 2007; Staedler and Endress, 2009). Recent comprehensive studies of the origin of the ancestral flower in angiosperms suggest that it consisted of a combination of whorled and spirally arranged organs (Endress and Doyle, 2009; Sauquet *et al.*, 2017). The perianth and androecium of the 'reconstructed' ancestral flower were likely whorled with three organs per whorl. In contrast, the carpels were likely arranged in a spiral pattern (Sauquet *et al.*, 2017). In a follow-up publication, Sokoloff *et al.* (2018) suggested that ancestral flowers were either 'entirely whorled (up to the gynoecium) or entirely spiral'. These conclusions confirm the potential for labile expression of phyllotactic patterns in flowers (Reyes *et al.*, 2018) as well as the existence of developmental constraints supported by the lack of examples of changes in phyllotaxis from androecium to gynoecium (Sokoloff *et al.*, 2018).

The observation of phyllotactic patterns also extends beyond land plants, and in some cases, as outlined in Peaucelle and Couder (2016) and Linardic and Braybrook (2017), these patterns closely resemble those of angiosperms although the structure and functioning of the apex are not the same. *Sargassum miticum*, a brown alga with a highly organised thallus, displays a spiral Fibonacci arrangement of lateral appendages (Fig. 6.1D–E); other examples of Sargassaceae have also been described as having common spiral phyllotactic patterns. In *Sargassum*, Peaucelle and Couder (2016) noted the equal probability of clockwise and counterclockwise winding of spirals as well as a distribution of divergence angles around the ideal angle of 137.5° similar to those of *Arabidopsis*. Additionally, their detailed morphological comparative study of *Arabidopsis* and *Sargassum* revealed large differences in the meristematic regions in both species. In *Sargassum*, the meristematic region forms a pit-like structure surrounded by smaller pits with an outer bulging part arranged in a phyllotactic pattern. These pits are responsible for the organised pattern formed by the lateral appendages. In their developmental study of the same species, Linardic and Braybrook (2017) noted that the phyllotactic pattern and the apical cell division pattern are not linked and are likely controlled separately. Additionally, they confirmed a self-organisation pattern when they observed the formation of a new central meristem after experimentally destroying the apical cell. The *Sargassum* system reaffirms the relevance of a self-organising process in understanding the foundational aspects of phyllotaxis where similar outcomes or patterns are expressed despite different underlying meristematic structures (Linardic and Braybrook, 2017). In this context, a focus on organising principles or processes such as growth rate, duration, and distribution among others (*sensu* Sattler, 1988, 1992) governing the generation of form, instead of specific genes, may facilitate a holistic understanding of phyllotaxis (Medina, 2010). From this perspective, genetic pathways or cascades provide the stability to forms already organised by inherent properties; this means that genetic programs represent a secondary effect or by-product of self-organisation. This 'internalist' perspective described by Medina (2010) is in contrast to an 'externalist' perspective in which organisation would be a secondary effect or by-product of genetic programs subjected to sorting or chance. These two perspectives are represented to different extents in developmental evolutionary biology, also known as EvoDevo.

Relationship between Physics and Biochemistry

Both physical constraints and biological processes play a role in plant development. This is why phyllotaxis cannot be fully understood without a comparative analysis between empirical data and theoretical models. The history of phyllotaxis can be thought of in terms of a pendulum oscillating across time between theory and experiment (Jean, 1994; Adler *et al.*, 1997; Korn, 2008). In addition to the 'pendulum' effect between theoretical and empirical studies, which moves the field forward, the next step is to combine different approaches (i.e. considering mechanical and biochemical models together as done by Newell *et al.* [2008a,b]) to increase the predictability or explanatory power of models.

In addition to the biochemical and molecular processes, there is also evidence that mechanical signalling is intimately involved in SAM growth and organogenesis (Kierzkowski *et al.*, 2012). Mechanical signals are not just a passive consequence of gene action; they can also have a feedback effect on morphogenesis (Jackson *et al.*, 2019; Kierzkowski and Routier-Kierzkowska, 2019). Mechanical forces and biochemical processes are closely linked (Newell *et al.*, 2008a,b; Sassi and Vernoux, 2013; Hamant, 2017) (Fig. 6.2). This aspect of morphogenesis is covered in greater detail in Chapter 5 on "Biophysical Aspects of Phyllotaxis" and summarised comprehensively in Fig. 6.2 (Sassi and Vernoux, 2013).

Growth Modes

In phyllotaxis, open and closed systems are observed from a developmental viewpoint (Barabé, 1995). In open systems such as those of *Arabidopsis* or *Anagalis*, lateral elements are produced successively at the periphery of the SAM and the central zone remains devoid of primordia during the growth of the structure.

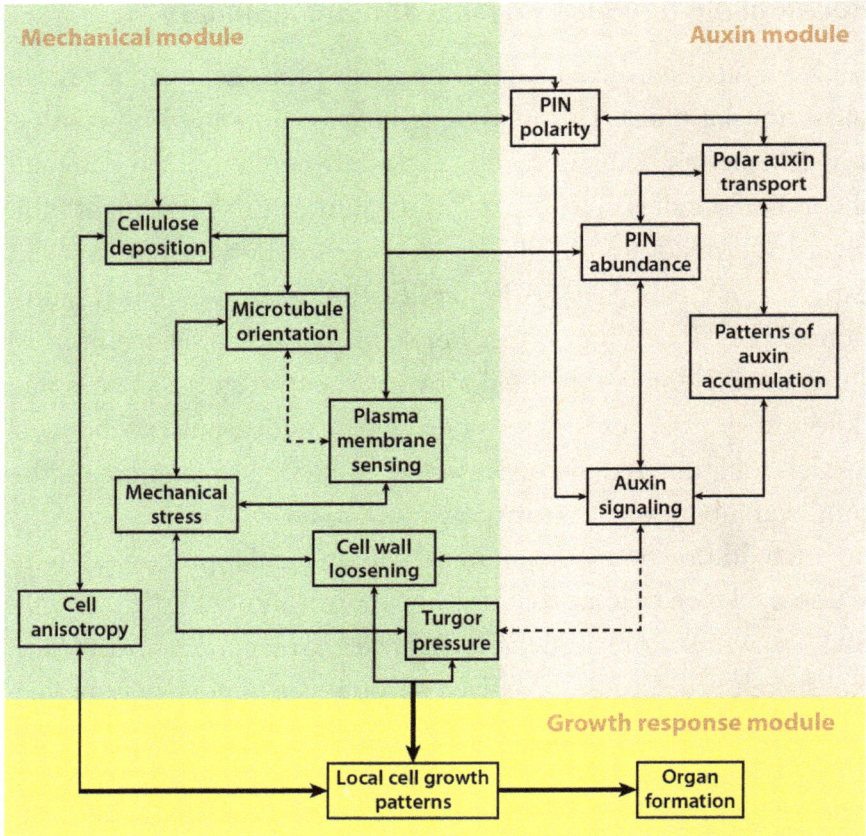

Fig. 6.2.   Crosstalk between auxin and mechanical signalling in the regulation of morphogenesis at the SAM. Auxin-induced cell-wall loosening is thought to cause mechanical stress that in turn feeds back on PIN1-mediated auxin transport to amplify local growth responses (see text for details). Dashed lines represent putative interactions: (1) auxin might also regulate the turgor pressure of the SAM cells, probably by modulating aquaporin activity, as observed recently in the lateral root formation process (Péret *et al.*, 2012); (2) the mechanical stresses that determine microtubule orientation (Hamant *et al.*, 2008; Heisler *et al.*, 2010; Uyttewaal *et al.*, 2012) might be sensed via the plasma membrane, as proposed recently for mechanical stress-driven changes in PIN1 intracellular distribution (Nakayama *et al.*, 2012). (From Sasi and Vernoux, 2013; reprinted with permission from Oxford University Press.)

This type of system is common in vegetative shoots and articulated inflorescences. Alternatively, in closed systems, which characterise flowers and compact inflorescences such as the spadix of Araceae (e.g. *Philodendron*) and the capitula of Asteraceae (e.g. *Helianthus*), the floral primordia will progressively cover all meristematic surfaces (Barabé *et al.*, 2000; Dosio *et al.*, 2006). Further, more than one developmental pathway may be involved in closed inflorescences depending on the taxonomic group (Bull-Hereñu and Claßen-Bockhoff, 2011). Then, are the phyllotactic processes and rules involved in open systems consisting of fewer primordia at the SAM level (e.g. *Arabidopsis*) the same as those in closed systems with many primordia (e.g. *Helianthus*)? Hotton *et al.* (2006) used the cone morphology of *Picea* to theoretically show how two growth modes, a constant Γ value (representing the new primordium radius over the meristem radius) and a constant value of the diameter of floral primordia (over a constant diameter of the meristem) with the same initial conditions, can lead to two different phyllotactic patterns. The constant Γ value would characterise indeterminate meristems (open systems) and intermediate stages of the development of capitula in Asteraceae, in which primordia never filled the meristem. However, when the primordium and meristem radii are constant, primordia fill up the centre of the meristem as is the case in many final development stages of cones of *Picea* and flowers (Fig. 6.3B).

Developmental rules acting in geometrical models can be considered a consequence of the molecular and cellular processes controlling primordium initiation (Hotton *et al.*, 2006). Then, can the phyllotactic patterns emerging in both growth modes be explained by the same iteration rules and the same biological processes? Can we explain this phenomenon by the same rules that explain the appearance of phyllotactic patterns in *Arabidopsis*?

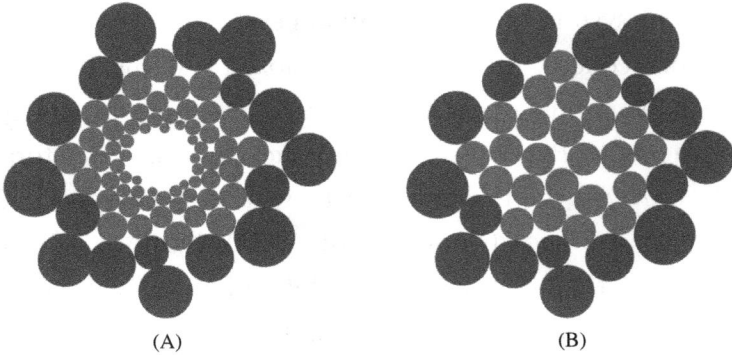

(A)   (B)

Fig. 6.3.   Two modes of growth with the same initial condition (in dark grey) derived from a micrograph of *Picea abies*. (A) The constant **ratio** model ($\Gamma = 0.176$ on a unit radius meristem) yields a pattern which, although not a lattice, does exhibit families of spirals that are parastichy like, with parastichy numbers (8, 13), which are the parastichy numbers of the original pattern. (B) The same initial condition yields a very different pattern with the constant **radius** model ($r = 0.176$). (From Hotton *et al.*, 2006; reprinted with permission from Springer Nature.)

Comparison between Models

The generation of the same types of phyllotactic patterns raises the question of the determination of common denominators between models. In this regard, Adler (1998) reported the suggestion of the physicist Phillip Morrison. Namely, Adler's (1974), Jean's (1980), Levitov's (1991a), and Douady and Couder's (1992) models, although based on three different assumptions (minimum distance between primordia, entropy, and repulsive energy), are mathematically equivalent. However, this remains a conjecture. Jean (1998b) compared phyllotactic patterns (number of parastichies) generated by his entropy-based model with those obtained using Cumming and Strickland's model (1998), in which pattern generation is based on the Helmholtz and Laplace equations of mathematical physics.

Guerreiro and Rothen (1998) analysed the conditions for the occurrence of the golden angle (137°) from a very general viewpoint. In their model, which is based on an inhibitory morphogen, they used an approach equivalent to that of Marzec and Kappraff (1983). They showed that the mathematical properties of the divergence angle are not the consequence of a precise growth mechanism but merely a statistical consequence of the distribution of inhibition. They also highlighted two important theoretical aspects regarding the nature of phyllotactic patterns and the selection of phyllotactic divergence angles. They concluded that 'regular phyllotaxis emerges as a consequence of quasi-geometrical rules almost independently of the details of the model considered' (p. 598) and that 'the observed selection of the divergences is better and more simply understood as a phenomenon of statistical (discontinuous selection) rather than as a phenomenon of progressive selection due to the continuous variation of the parameters governing the meristematic growing process, as suggested by Douady and Couder (1992) for example' (p. 598). More recently, Reick (2012, 2015) explained the occurrence of particular divergence angles in phyllotactic systems on the basis of self-similarity and scaling processes that appear in bifurcation diagrams resulting from dynamical phyllotaxis models like those of Thornley (1975) and of Douady and Couder (1992). By renormalising the theoretical inhibitory potential governing the mutual inhibition of primordia, Reick (2015) showed a self-similarity of the bifurcation structure present in the models. The universal appearance of certain phyllotactic patterns is independent of biological processes involved in SAM morphogenesis. Reick's comparison method (2012, 2015) appears promising in the search for universal structural characteristics common to a certain class of dynamical phyllotactic models. In their own approach to phyllotaxis, Shipman and Newell (2005) also questioned the extent to which Douady and Couder's dynamic rules (1996a) are contained in their mechanical model.

Newell *et al.* (2008a) combined a biochemical (Jönsson *et al.*, 2006) and a biophysical model (Shipman and Newell, 2005) by using continuous differential equations. Further, they provided insights for advancing phyllotaxis by questioning the extent to which the repulsive energy approach to phyllotaxis central to the studies of Levitov (1991a,b), Douady and Couder (1996a,b), Atela *et al.* (2002), and Hotton *et al.* (2006) can be linked to gradient or non-gradient PDEs describing biochemical (e.g. Jönsson *et al.* (2006), Newell *et al.*, 2008a,b), or biomechanical models (e.g. Shipman and Newell, 2005). Because (differential) equations derived from various proposed mechanisms share a similar form, understanding this form can give insights into fundamental mathematical principles of phyllotaxis and also help us to understand what a model can and cannot tell us (Shipman *et al.*, 2011). If a model produces a phyllotactic pattern, it may still be totally incorrect because of the common structure of pattern-forming systems (Pennybacker *et al.*, 2015).

All phyllotactic models from the simplest to the most complex can generate similar phyllotactic patterns in terms of number of parastichies and divergence angle. Shipman and Newell (2005) stated that different microscopic processes proposed to explain phyllotactic patterns can produced similar macroscopic patterns. Newell *et al.* (2008a,b) described biochemical models and biophysical models by the same quadratic equation. This would indicate that the models converge not because of the biological explanation involved but because of the equations used to describe the generation of phyllotactic patterns. Therefore, if different models can predict the same phyllotactic patterns, how can we prove the value of a model?

During the appearance of phyllotactic patterns, two different phases can be distinguished. The first one corresponds to the development of a primordium on the SAM surface. The explanation of the appearance of a primordium can be based, for example, on biochemical or physical processes or both. The second phase corresponds to the reiteration of the phyllotactic pattern itself. For a phyllotactic pattern to emerge, periodicity in the form of a reiterative process must be established. It is precisely because we have to deal with a reiterative process that there is a convergence between different models. The convergence between models may be a consequence of the iteration rules involved in the propagation of the phyllotactic patterns during plant growth or more generally over time.

The common point between different models could be a parameter analogous to parameter Γ that considers the SAM size (or generative region) in relation to an element linked directly or indirectly to the primordium size or the plastochrone ratio. This could be represented by the following ratios: the ratio of SAM diameter to leaf diameter in geometrical models; the ratio of SAM diameter to the size of inhibition field in chemical models; the ratio of SAM diameter to wavelength in biophysical models; and size of central zone in biochemical models.

Parameter Γ (Link between Models)

Gola and Banasiak (2016) emphasised the importance of considering the large diversity of SAM morphology in phyllotactic analysis. Nearly all theoretical models simulating the emergence of phyllotactic patterns deal with a SAM where the leaf diameter is smaller than the SAM diameter. For example, in Shipman and Newell's model (2005), deformations on the SAM surface can be described as a linear combination of a periodic function if the ratio of the width and circumference of the generative region to the intrinsic wavelength linked to the buckling of the tunica is large. Is this model applicable when the apex is very small in relation to the size of foliar primordia, as seen in *Begonia* and many monocotyledons, particularly in palms?

In theoretical models, this Γ value is greater in distichous and spirodistichous systems than in spiral systems. However, this is not always the case. For example, in *Begonia*, we observe a typical spirodistichous system where the divergence angle varies between 165° and 180° (Barabé *et al.*, 2007). In *Euterpe*, there is a spiral system characterised by an angle varying from 116° to 157° (Barabé *et al.*, 2010). However, in both cases, the SAM diameter is very small in relation to the leaf diameter, which translates to similar Γ values, which normally characterise spirodistichous systems (Barabé *et al.*, 2010). In the case of *Begonia*, the SAM volume is so small in relation to that of the foliar primordium that it leads us to ask whether the emergence of the phyllotactic pattern is governed by the SAM of the two adjacent facing leaf primordia. It would be interesting to develop a model that considers the form and size of the apical meristem in relation to the primordium size.

How can we explain the high Γ value in both cases but a spiral pattern in *Euterpe* and a distichous pattern in *Begonia*? Is the Γ value only a consequence of the mode of development or a geometrical control parameter as advocated in many models? If it is a consequence of the mode of development, then the control of phyllotaxis remains at the microscopic level. However, as indicated previously (see evolution section), the structure of the apex is different in brown alga and flowering plants but both groups can generate spiral phyllotactic patterns. Does this mean that the overall structure of phyllotactic patterns is not dependent on cellular organisation?

Developmental Constraints, Genetics, and Self-Organisation

Phyllotaxis is a developmental constraint but not an adaptive one because the efficiency of a phyllotactic pattern in terms of light capture can be offset by a particular stem morphology or leaf orientation (Amundson, 1994). Phyllotaxis is not a primary adaptive character; the adaptive character is the overall plant architecture or structure. However, phyllotaxis can be linked to the mode of growth. For example, plants with a rosette of leaves, such as *Achmea*, tend to show spiral phyllotaxis. Other plants such as climbing vines have leaf blades oriented parallel to the ground to maximise light capture. It would be interesting to study the relationship between a plant's phyllotaxis and its growth mode and habitat.

Seilacher (1991) and De Renzi (1997) noted that in organisms, the genome creates conditions under which a proper 'development environment' is available and turns on autonomous morphogenetic mechanisms, resulting in structures that become self-organised as opposed to coding for these structures directly. When searching for analogies in the physical world, Seilacher (1991, p. 7) stated that 'We should remember, however, that similar outcomes may refer only to the rules and not to the chemical or mechanical causations'. Further, Douady and Couder (1998, p. 567) noted that 'The genetic system does not fix the phyllotaxy directly, rather it determines the system that will generate it'.

Douady and Couder (1996a) noted that some genetic factors are involved in the appearance of phyllotactic patterns. For example, the ratio of the primordium diameter to the SAM diameter is determined by the genetic characteristics that constitute the boundary conditions of the system. What, then, is the relative role of genetics and self-organisation in the phyllotactic organisation of an individual plant? The genetic system lays down the foundation (constraints) on which development will take place. For example, in molecular models that highlight the activity of certain genes, these genes are not necessarily specific in that they modulate specific phyllotactic parameters. Rather, genes are linked to specific developmental processes that ultimately have an impact on the biological constraints within which phyllotactic patterns develop. This is a subtle but important distinction.

The 'genetic' perspective can also be viewed from a more hierarchical viewpoint that emphasises the causal chain of steps between the plant's genome and its ultimate morphology by using mathematical models (Green, 1996, 1999). For example, at the whole plant level, biological and physical constraints (as well as genetic factors) affect the overall structure of a plant. At the SAM level, other factors (both genetic and structural) may affect the type of pattern that will emerge. Depending on the level of observation (cell vs. SAM), different limitations (constraints) exist and different genes and their involvement in developmental processes are in operation.

Backmann (1983) used the term 'canalisation', coined by Waddington (1940), instead of 'differentiation' in the development of complex organs. This term indicates that the development of an organ with a high organisational level can be viewed as a canalisation to a limited choice of possible forms. Bachmann (1983) explained numerical canalisation in plant phyllotaxis, such as the frequent occurrence of Fibonacci series, by a self-organisation process at the SAM level. Phyllotactic patterns emerge from a few simple epigenetic rules under the constraint of species-specific boundary conditions. He noted that 'the genetic information can only set the boundary conditions under which the rules are elaborated' (p. 191). If the boundary conditions are modified, a new phyllotactic pattern will appear. This phenomenon was analysed in detail in *Microseris capitula* (Battjes *et al.*, 1993). In this canalisation towards a Fibonacci trajectory, the position of a primordium could be influenced by the position of older neighbouring primordia.

Genetic Constraints

The presence of a species-specific genetic constraint is well-illustrated by the case of *Thuja* (Yin *et al.*, 2011). In *Thuja occidentalis*, the typical phyllotactic pattern is opposite-decussate. This is a genetic characteristic of this species. However, three pathways to the decussate pattern are observed in seedlings: no pattern transition from the beginning, from tetracussate to decussate, and from tetracussate to tricussate to spiral to decussate (Fig. 6.4). This indicates that the decussate pattern is genetically fixed and represents a constraint on phyllotactic patterns that can be formed. In this particular system, the tricussate and spiral patterns are unstable patterns that will eventually transition to the stable pattern characteristic of the species. Micro instability in the biological processes involved in the SAM of seedlings may explain the presence of intermediate patterns.

Main Challenges in Phyllotaxis

Jean (1998a) identified six main challenges in phyllotaxis. Of these, two were immediate challenges that needed further

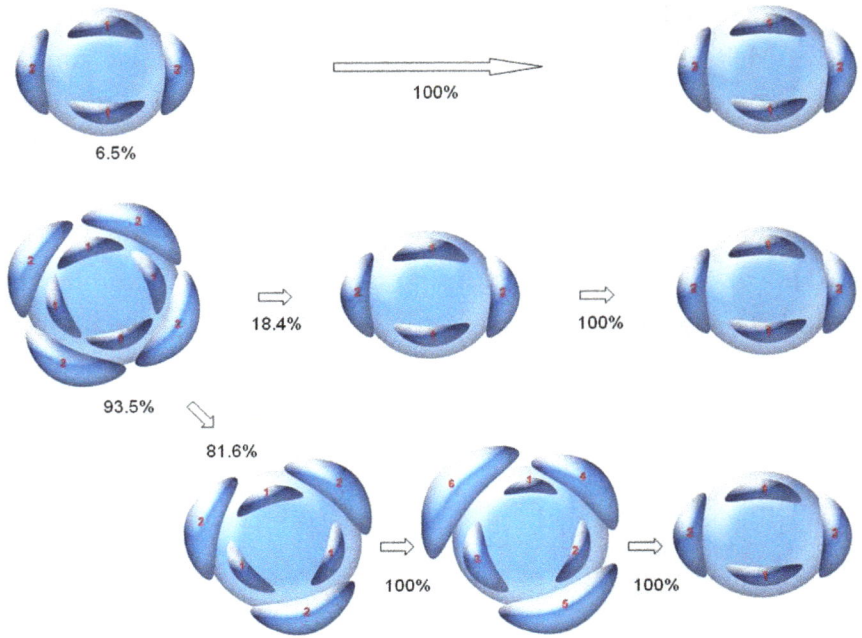

Fig. 6.4. Diagrammatic representation of observed phyllotactic pattern transitions. Percentage values refer to the proportion of apices involved in the particular transition. (From Yin *et al.*, 2011; reprinted with permission from Canadian Science Publishing.)

study: pattern recognition and pattern generation. The 20 years since have seen two general approaches to phyllotactic studies: (1) the elaboration of models based on biochemical, biomechanical, or chemical principles and (2) the analysis of the emergence of phyllotactic patterns in terms of geometrical and numerical properties behind equations describing the models. Most studies discussed in this paper focussed on the first approach. Sophisticated and accurate phyllotaxis models based on empirical evidence have been developed and invoke biological processes that occur at the cellular or the SAM level. It would be interesting to know how the models compare in their predictive capacity and how they can be used synergistically. Here, synergy implies that there are aspects of different models that, when combined, could afford greater predictive capacity, that is, explain a greater diversity of phyllotactic patterns. In recent years, the second approach has been the object of many analyses showing the mathematical similarities between different dynamic models and geometrical models. This aspect, dealing more with mathematics than biology, has not been covered in detail in the present book. However, this subject could certainly be developed further by theoreticians in a comprehensive book.

The third challenge is the phyletic origin(s) of phyllotactic patterns. This aspect of phyllotaxis has not been studied in as much detail as pattern recognition or generation. Available information suggests that although a wide variety of morphologies in terms of meristematic regions are involved in the manifestation of phyllotactic patterns, universal self-organising processes appear to be foundational to the elaboration of common patterns regardless of the different underlying structural constraints across taxa. However, no phylogenetic analysis shows how phyllotactic patterns may vary among taxa. For example, are the frequency and evolution of phyllotactic patterns the same in monocotyledons and dicotyledons considering that the SAM size and morphology are different in both groups?

The fourth challenge is connected in part to the third challenge and deals with the functional aspects of patterns and their relevance to plants in terms of selective advantage or adaptability. This challenge highlights the practical aspects of phyllotaxis in a more ecological context and the potential to further develop quantitative aspects of the costs and benefits of phyllotactic patterns. Thus far, very few studies (Valladares and Brites, 2004) have addressed the question of the ecological perspective of phyllotaxis. Although phyllotaxis does not appear to be an adaptive character, it is certainly linked to the growth mode and habitat of a plant. This question remains open and needs further study.

The broadest challenge is explaining why the types of 'geometric' patterns that are expressed in plants are more common in this group than in other organisms or even in nature as a whole. The unique development modes of the SAM in various taxa, often termed indeterminate growth, are likely linked to the broad array of patterns that are possible in plants and plant-like organisms. In recent years, the further characterisation of the SAM and its role(s) in the elaboration of phyllotactic patterns have shown how a relatively rigid brick-like assemblage of cells can lead to the formation and iteration of regular and stable phyllotaxies.

Plant growth can be viewed as the iteration of growth units corresponding to different morphological entities such as metamers or phytons (Rutishauser and Sattler, 1985). Priestly and Scott (1933) addressed the anatomical relationship between shoot units and phyllotaxis a long time ago. In a more abstract sense, the growth unit corresponds to Jean's notion of 'gnomon' (1998a). Geometrically, a gnomon is an elementary form repeating itself a given number of times. Gnomic growth corresponds to the iterative addition of this elementary form. Jean (1994, 1998a) considered that this rhythmic growth mode will produce a hierarchical structure that is mathematically linked to the appearance of phyllotactic patterns. However, this theoretical representation may only be appropriate to describe the indeterminate growth of a stem (open system) where the growth units are generally easy to identify; it becomes more problematic in structures with determinate growth (closed systems), such as flowers or compact inflorescences (spadix, capitulum).

The sixth and final challenge is to produce a comprehensive history of phyllotaxis. Jean (1998a) mentioned that many elements of a historical account of the discipline already exist. Although the history of phyllotaxis is not one of the aims of our book, we nonetheless present a comprehensive timeline of developments that include recent advances in the field and provide the opportunity to assemble a more complete snapshot of the discipline that could ultimately benefit the development of this subject. We know that the history of phyllotaxis involves a continuous cross-disciplinary overlap between observational data, experimental studies, and theoretical models (Adler *et al.*, 1997; Jean and Barabé, 1998b). Consequently, we see a sequence of models going from descriptive geometrical models to recent theoretical models incorporating biochemical and biomechanical processes. This sequence has led to more robust and comprehensive models that are prominent now. However, it would be important to historically and philosophically analyse how different concepts, perspective, and methods have contributed to our understanding of phyllotaxis.

General Conclusion

The study of phyllotaxis may turn out to be a strong example of the value of multidisciplinary approaches to understanding pattern formation in plants from their establishment to their manifestation.

Today, with the sophistication of numerical simulations, development of precise quantitative methods, and progress of molecular approaches, phyllotaxis may advance as a research discipline. We can now generate patterns that can be viewed as three-dimensionally and manipulate various parameters easily (Routier-Kierkowska and Runions, 2018). For example, a recent technological development called automated confocal micro-extensometer (ACME) shows tremendous promise in terms of accurately measuring the mechanical properties of plant cells and tissues (Robinson *et al.*, 2017). The ACME device is a small version of an extensometer that relies on confocal imaging to measure strain, and it can be used on living tissues. As proof of concept, Robinson *et al.* (2017) showed that gibberellic acid (GA) increased the elasticity of cell walls in the hypocotyls of *Arabidopsis*. Another method that shows promise is the MorphoGraphX platform developed to quantify morphogenesis at multiple time points. It will likely prove useful to study the reiteration of phyllotactic patterns (Barbier de Reuille *et al.* (2015). This type of empirical information can prove to be extremely useful in the development of future models to explain morphogenetic events and, more specifically, phyllotactic patterning in plants. The phylogenetic and developmental relationships between the SAM of monocotyledons and dicotyledons should be further explored with respect to molecular biology and models to further elucidate more universal processes linked to the manifestation of phyllotactic patterns.

From a theoretical viewpoint, we have observed that different models (i.e. chemical, mechanical, biochemical) can lead to the emergence of the same phyllotactic patterns. Consequently, it may be crucial to determine if there are common underlying mathematical principles at work. This may help to determine if phyllotaxis results from the intrinsic structure of the equations or the biological or physical processes involved in these models.

Phyllotaxis is a good example of the type of approaches that can be addressed in theoretical biology. The reciprocal flow of data and information between theory and experimental studies in the history of phyllotaxis shows how theoretical models can be used to assess quantitative relationships between different patterns and how empirical data can stimulate the development of new theoretical approaches.

Glossary

Note: Readers should also refer to Jean 1994 (Phyllotaxis), which contains a comprehensive glossary of terms related to phyllotaxis.

Activator: One of two substances involved in a reaction–diffusion system which consists of two chemical substances: an activator and inhibitor.

Airy stress function: Equation describing the deformation of an elastic body subject to forces on the boundary in two dimensions.

Angle of divergence (also divergence angle): In a planar representation of the shoot apical meristem (SAM), the smallest angle between two successive leaves using the SAM centre as the reference point.

Annular region (also annular generative region): Term used in biomechanical models to designate the zone of the SAM where primordia are initiated (Shipman and Newell, 2005).

Anodic: Along a spiral pattern of leaves, the clockwise (ascending) sequence of organs from older to younger structures.

Anticlinal: Orientation of cell wall perpendicular to the surface of a structure.

Apical meristem (also apex or SAM): Region of active cell division and morphogenesis at the tip of a shoot, floral meristem, or root.

Apoplast: Extracellular space (outside the plasma membrane) where materials can diffuse freely within and between cellulosic cell walls.

Aquaporins: Proteins located on the cellular membrane that form pores through which water can flow into and out of cells (also referred to as water channels).

Auxin: Phytohormone involved in a variety of morphogenetic growth processes.

Axial sympodial arrangement: Mode of development where the main shoot axis terminates and side branches take over the main growth of the plant.

Axillary meristem: Meristem that is located at the junction of a leaf and the stem and that gives rise to a branch.

Azimuthal: In an 'azimuthal' map projection of a SAM, all points on the map are at proportionately correct distances from the centre point.

Biastrepsis: Particular torsion of the stem in various herbaceous plants with whorled or decussate phyllotaxis.

Bifurcation diagram: In phyllotaxis, a bifurcation diagram shows how the values of the divergence angle (d) and number of parastichies (m, n) can change in relation to the continuous variation of a control parameter (e.g. the rise or the plastochrone ratio) in the system. For certain values of the control parameter, a new bifurcation appears in the diagram and a branch is added. The bifurcation diagram is also called a phyllotactic tree, given its resemblance to a tree.

Bijugy: System consisting of two parallel ontogenetic helices (also referred to as genetic spiral) along which the generation of primordia occurs, resulting in complex phyllotactic patterns.

Bract: Leaf-like structure found at the base of some inflorescences or flowers.

Bulk entropy: A concept developed by Jean (1994) in which the entropy of a phyllotactic system is based on the growth rhythm of the hierarchy of that system.

Canalisation: The directional and sequential organisation of structures during morphogenesis. From a biochemical perspective, it represents the consolidation of flow/flux of a morphogen in a specific direction.

Canonical angles: In the context of phyllotaxis, these correspond to specific patterns ultimately represented by a divergence angle of 137.5°.

Capitula: Morphological term describing the enlarged discoid inflorescence meristem typically found in Asteraceae.

Cathodic: Along a spiral pattern of leaves, the counterclockwise (descending) sequence of organs from younger to older structures.

Central zone: Area in the central portion of a SAM where cells divide slowly.

Chemiosmotic transport: Movement of ions across a semi-permeable cell membrane down an electrochemical gradient.

Circumferential curvature: Measure of the curvature along the circumference of the generative region on the SAM in the context of biomechanical models.

Conicity: Characterisation of cone shapes.

Conspicuous parastichies (according to Jean, 1994): A visible opposed parastichy pair for which the intersection angle of opposed spirals is closer to 90° than the intersection angle of any other visible opposed parastichy pair of the system.

Contact parastichies: Phyllotactic spirals or helices along which the bases of leaf primordia are in contact before intercalary growth occurs.

Continuous transition: Symmetrical expansions or contractions of the shoot perimeter resulting in uniform changes in the quantitative relationships between the primordia and the SAM.

Corpus: In angiosperm SAM, a group of cells located underneath the tunica layers that divide in no particular plane.

Cytokinin: Phytohormone involved primarily in cell division, growth, and differentiation in plant roots and shoots.

Decussate (and opposite decussate): Leaf initiation in pairs at the same level of insertion, which alternate by 90° between successive whorls.

Determinate: Growth of a structure to a specific size and shape.

Discontinuous transition: Abrupt change in phyllotactic pattern that occurs via asymmetrical expansions or contractions in localised sectors of the shoot system and results in the addition or loss of a parastichy in a family.

Distichous: Alternate arrangement of single leaves in two opposite rows (orthostichies) at 180°.

Divergence angle: See angle of divergence.

Efflux: Direction of flow away from a source.

Endosome: Membrane-bound compartment inside eukaryotic cells.

Entropy: From a general viewpoint, a physical measure of the degree of order in a system.

Epigenetic: Heritable changes in gene function that do not involve changes in the DNA sequence.

Expansin: Non-enzymatic protein found in the plant cell wall and involved in developmental processes where cell wall loosening occurs.

Fibonacci numbers (also Fibonacci series, Fibonacci spiral): Any number in the Fibonacci sequence of integers (1, 1, 2, 3, 5, 8, 13, …) that is the sum of the two previous numbers in the sequence.

Fick's Laws: Mathematical descriptions of the diffusion process for different diffusion coefficients. The first law relates the diffusion flux to the concentration gradient. The second law predicts how diffusion causes the concentration to change with time.

Floral meristem: Meristematic structure that will produce the floral organs of a flower.

Flux lattice: Energy which passes through a surface or substance to create a hierarchical structure similar to phyllotactic patterns.

Foliar primordium: Leaf at an early stage of initiation.

Folioid: Leaf-like element.

Fundamental theorem of phyllotaxis: Mathematical relationship that associates divergence angle with the number of parastichies; developed by Adler (1974) and Jean (1988).

Gaussian curvature: Gaussian curvature of a surface is an intrinsic measure of curvature independent of the system of coordinates used to describe it.

Gaussian function: An exponential function of the form $\exp(-x^2)$ represented by a characteristic symmetrical bell curve.

Generative region: See annular region.

Generative spiral (also fundamental spiral, ontogenetic spiral): Hypothetical curve linking the initiation of successive leaves on a shoot system.

Gibberellic acid (also gibberellin): Phytohormone regulating various developmental processes, including stem elongation, germination, dormancy, flowering, flower development, and leaf and fruit senescence.

Gnomic growth: Iterative addition of a gnomon.

Gnomon: An elementary form repeating itself a given number of times.

Golden angle: Divergence angle associated with the limit of the Fibonacci series (137.5°).

Histogenetic centre: In some descriptions of phyllotaxis, this is referred to as the zone of initiation of primordia.

Hofmeister rule: Primordia originate the furthest away from the base of existing leaves.

Homeobox: DNA sequence, around 180 base pairs long, found within genes that are involved in the regulation of patterns of anatomical development (morphogenesis).

Hyperbolic tessellations: In hyperbolic geometry (non-Euclidean geometry), a uniform hyperbolic tilling is an edge-to-edge filling of the hyperbolic plan which consists of regular polygons and where all the vertices are equivalent under the symmetries of the figure.

Indeterminate: Continuous growth of a structure without a specific final stage. For example, the SAM is typically an indeterminate structure, unless it transitions to a flower or inflorescence.

Indole Acetic Acid (IAA): Phytohormone of the auxin class involved in inducing cell elongation and cell division; it also serves as a signalling molecule necessary for development of plant organs and coordination of growth.

Inflorescence: A collection of flowers arranged in a group or cluster (i.e. sunflower capitulum).

Influx: Direction of flow towards a specific location.

Inhibitor: See activator.

Initium: Earliest stage of primordium initiation.

KNOX gene: Genes belonging to this family essentially maintain stem cells in an undifferentiated state.

Lamina: A leaf blade.

Leaf arc (also leaf insertion angle, leaf arc angle): The angle between the edges of a primordium in relation to the SAM centre in a planar representation.

Leaf primordium: Early stage of leaf development.

Markov chain: A stochastic model describing a sequence of possible events in which the probability of each event depends only on the state attained in the previous event.

Meristem: Region of active cell division and morphogenesis in plants.

Merophyte: Segment that is derived from an apical cell (as in ferns).

Michaelis-Menten function: Describes the kinetics of a reaction catalysed by an enzyme, relating the rate of formation of product to the concentration of a substrate.

Monostichy: The alignment of all leaves along a single sector (orthostichy), found in various ferns and a few flowering plants.

Morphogen: A theoretical chemical substance whose nonuniform distribution regulates the pattern of tissue development in the process of morphogenesis or pattern formation.

Multijugate: Multiple number of parallel genetic spirals along which the generation of primordia occurs. Bijugy is an example of multijugy in which there are two parallel helices.

Negentropy: The opposite of entropy. In a biological context, the negentropy of a living system counteracts the tendency towards entropy. In the context of phyllotaxis, it means that phyllotactic organisation becomes more orderly with increasing negentropy.

Node: Point of attachment of leaves along a stem.

Opposite-decussate: See decussate.

Organogenesis: Morphological development (morphogenesis) of organs.

Orthostichy: A straight line obtained by joining two primordia that are approximately superposed with respect to the SAM. According to Jean (1994), an orthostichy is a form of parastichy.

Oscillation distichy: Distichous phyllotactic pattern where consecutive pairs of leaves shift in position (back and forth) in relation to each other.

Parastichy: Any type of spirals (or helices) linking primordia in a phyllotactic system.

Pectin: A complex set of polysaccharides contained in the primary cell walls of terrestrial plants.

Pendulum symmetry (*Pendelsymmetrie*): Development patterns with some kind of left/right oscillation of symmetry.

Pentamerous: Grouping of five organs or structures.

Peripheral zone: Zone located at the SAM periphery where primordium initiation typically occurs.

Perturbed pattern: Dynamic phyllotactic pattern referred to as chaotic, erratic, scattered, irregular, or abnormal.

Phyllotactic diagram: See bifurcation diagram.

Phyllotactic lattice: A repeating arrangement of points on a planar surface depicting a phyllotactic pattern.

Phyllotactic tiling planform: Geometrical figure (e.g. ridges, irregular hexagons, parallelograms) formed by the bases of leaves (tiles) on a stem or SAM.

Phyllotactic tree: See bifurcation diagram.

Phyllotaxis: Process of description, characterisation, and generation of patterns made by similar organs (e.g. leaves, floral parts) on plants.

Phyllotaxy: The disposition or arrangement of leaves on a shoot system.

Phytomer: Functional unit of a plant, continually produced by shoot meristems; each unit consists of a node to which a leaf is attached, a subtending internode, and an axillary bud at the base of the leaf.

PIN protein: Integral membrane protein that transports the anionic form of the phytohormone auxin across membranes; PIN1 is an example of this type of protein.

Plastochrone: The time interval between the initiation of two successive primordia or whorls of primordia on the SAM.

Plastochrone ratio: The ratio of the distance of two successive primordia from the SAM centre.

Poisson process: A type of random mathematical object that consists of points randomly located on a mathematical space in the context of probabilities and statistics.

Poisson's ratio: The negative of the ratio of transverse strain to axial strain. It is used to characterise the contraction of matter perpendicular to the force that is applied.

Primitive: A biological characteristic that does not differ from its ancestral sources.

Primordium: Early stage of development of a leaf, floral organ, flower, or inflorescence.

Pseudo-decussate: A spiral phyllotactic system that superficially looks like an opposed-decussate system.

Radial curvature: Measure of the curvature of the central portion of the generative region on the SAM in the context of biomechanical models.

Rise: The vertical internodal distance between two points (i.e. leaves) on the genetic spiral. The rise is related to the plastochrone ratio by the formula $r = \ln R/2\pi$ (Jean, 1994).

Shoot apical meristem (SAM): See apical meristem.

Snow's rule: A primordium forms in a specific area of the apex where there is adequate space for its initiation.

Spadix: Elongated cylindrical inflorescence typical of Aroids consisting of flowers arranged in compactly symmetrical rows.

Spiral: A three-dimensional curve that turns around an axis at a constant or continuously varying distance while moving parallel to the axis.

Spiro-decussate: Decussate phyllotactic system where successive pairs of leaves are inserted at an angle less than 90°, giving the impression of multiple spirals.

Spiro-distichous: Distichous phyllotactic system in which there is a regular twisting of the two orthostichies, resulting in a spiral stairway-like pattern.

Spirodistichy: See spiro-distichous.

Stochastic: Something that is determined randomly.

Tetramerous: Grouping of four organs or structures.

Theory of first available space: See Snow's rule.

Tiller buds: Stem produced by grass plants; refers to all shoots that grow after the initial parent shoot grows from a seed.

Transcription factor: Also referred to as a sequence-specific DNA-binding factor; protein that controls the rate of transcription of genetic information from DNA to messenger RNA by binding to a specific DNA sequence.

Tricussate: A verticil of structures (leaves, floral organs) consisting of three elements.

Tunica: The outermost one or two layers of the SAM of angiosperms where cell division is mainly anticlinal.

Turgor pressure: Also called hydrostatic pressure; pressure exerted by the osmotic flow of water in or out of a cell.

Vascullotaxy: Study of the organisation and evolution of vascular systems in plants.

Verticillate: Pattern of arrangement of leaves where two or more leaves form a ring at the same level (node) on the stem.

Visible opposed parastichies: The parastichies (m, n) that are formed when primordia are represented as points; the primordia are not necessarily touching each other but are found at the intersection of any two opposed parastichies.

Wavevector: A vector that helps to describe a wave. Its magnitude corresponds to the wave number or angular wave number of the wave, and its direction is ordinarily (not always) the direction of propagation of the wave.

Whorled: See verticillate.

Bibliography

Abley K, De Reuille PB, Strutt D, *et al.* (2013) An intracellular partitioning-based framework for tissue cell polarity in plants and animals. *Development* **140**: 2061–2074.

Abraham-Shrauner B and Pickard BG. (2011) A model for leaf initiation. Determination of phyllotaxis by waves in the generative circle. *Plant Signal Behav* **6**: 1755–1768.

Adler I. (1974) A model of contact pressure in phyllotaxis. *J Theor Biol* **45**: 1–79.

Adler I. (1977) The consequence of contact pressure in phyllotaxis. *J Theor Biol* **65**: 29–77.

Adler I. (1998) The role of continued fractions in phyllotaxis. *J Algebra* **205**: 227–243.

Adler I, Barabé D, and Jean RV. (1997) A history of the study of phyllotaxis. *Ann Bot* **80**: 231–244.

Airy H. (1873) On leaf-arrangement. *Proc R Soc Lon* **21**: 176–179.

Altesor A and Ezcurra E. (2003) Functional morphology and evolution of stem succulence in cacti. *J Arid Environ* **53**: 557–567.

Amundson R. (1994) Two concepts of constraint: adaptationism and the challenge from developmental biology. *Philos Sci* **61**: 556–578.

Atela P, Golé C, and Hotton S. (2002) A dynamical system for plant pattern formation: a rigorous analysis. *J Nonlinear Sci* **12**: 641–676.

Atela P, Golé C, and Hotton S. (2008) A dynamical system for plant pattern formation. http://maven.smith.edu/~phyllo/Assets/pdf/talk.pdf

Atela P. (2011) The geometric and dynamic essence of phyllotaxis. *Math Model Nat Phenom* **6**: 173–186.

Atlan H. (1972) *L'organisation biologique et la théorie de l'information.* Hermann, Paris.

Aulenback KR. (2013) A preliminary investigation into the study of vascullo-taxy (study nov.) based on selected extinct and extant taxa (Cupressaceae, *Ginkgo biloba*, *Populus deltoides*, Osmundaceae) and its bearing on stellar evolution within the seed plants. Published by the author.

Avery J. (2003) *Information Theory and Evolution*. World Scientific, Singapore.

Backmann K. (1983) Evolutionary genetics and the genetic control of morphogenesis in flowering plants. In: Hecht MK, Wallace B, Prance GT (eds), *Evolutionary Biology*. Springer, Boston, pp. 157–208. DOI:10.1007/978-14615-6971-8_5.

Bainbridge K, Cuyomarc'h S, Bayer E, *et al.* (2008) Auxin influx carriers stabilize phyllotactic patterning. *Genes Dev* **22**: 810–823.

Banasiak A. (2011) Putative dual pathway of auxin transport in organogenesis of *Arabidopsis*. *Planta* **233**: 49–61. DOI:10.1007/s00425-010-1280-0.

Bar M and Ori N. (2014) Leaf development and morphogenesis. *Development* **141**: 4219–4230.

Barabé D. (1995) Phyllotaxis: open and closed systems. *J Biol Syst* **3**: 917–927.

Barabé D. (2006) Stochastic approaches in phyllotaxis. *Can J Bot* **84**: 1675–1685.

Barabé D, Bourque L, Yin X, and Lacroix C. (2010) Phyllotaxis of the palm *Euterpe oleracea* Mart. At the level of the shoot apical meristem. *Botany* **88**: 528–536. DOI:10.1139/B10-010.

Barabé D, Brouillet L, and Bertrand C. (1992) Organogénie de la feuille du *Begonia radicans* Vellozo et du *Begonia scabrida* (Begoniaceae). *Can J Bot* **70**: 1107–1122.

Barabé D and Jean RV. (1996) The constraints of global form on phyllotactic organization: the case of *Symplocarpus* (Araceae). *J Theor Biol* **178**: 393–397.

Barabé D and Jeune B. (2004) The use of entropy to analyse phyllotactic mutants: a theoretical analysis. *Plant Cell* **16**: 804–806.

Barabé D and Jeune B. (2006) Complexity and information in regular and random phyllotactic patterns. *Riv Biol* **99**: 85–102.

Barabé D, Jeune B, and Lacroix C. (2009) Comparison between theoretical and empirical parameters in phyllotaxis: the case of Begonia. *Riv Biol* **102**: 157–164.

Barabé D and Lacroix C. (2008) Hierarchical developmental morphology: the case of the inflorescence of *Philodendron ornatum* (Araceae). *Int J Plant Sci* **169**: 1013–1022.

Barabé D, Lacroix C, and Jeune B. (2000) Development of the inflorescence and flower of *Philodendron fragrantissimum* (Araceae): a qualitative and a quantitative study. *Can J Bot* **78**: 557–576.

Barabé D, Lacroix C, and Jeune B. (2007) Following the initiation and development of individual leaf primordial at the level of shoot apical meristem (SAM): the case of distichous phyllotaxis in *Begonia*. *Ann Bot* **99**: 555–560.

Barabé D and Vieth J. (1990) La torsion de contrainte et le modèle phyllotaxique de Jean. *Can J Bot* **68**: 677–684.

Barbier de Reuille PB, Bohn-Courseau I, Ljung K, *et al.* (2006) Computer simulations reveal properties of the cell–cell signaling network at the shoot apex in *Arabidopsis*. *Proc Natl Acad Sci (PNAS)* **103**: 1627–1632.

Barbier de Reuille PB, Routier-Kierzkowska AL, Kierzkowski D, *et al.* (2015) MorphoGraphX: a platform for quantifying morphogenesis in 4D. *eLife* **4**: e05864. DOI:10.7554/eLife.05864.

Barlow PW. (1994) Rhythm, periodicity, and polarity as bases for morphogenesis in plants. *Biol Rev* **69**: 475–525.

Battjes J and Prusinkiewicz P. (1998) Modelling meristic characters of Asteraceen flowerheads. In: Jean RV, Barabé D (eds), *Symmetry in Plants*. World Scientific, Singapore, pp. 281–312.

Battjes J, Vischer NOE, and Bachmann C. (1993) Capitulum phyllotaxis and numerical canalization in *Microseris pygmaea* (Asreraceae: Lactuceae). *Amer J Bot* **80**: 419–428.

Bayer EM, Smith RS, Mandel T, *et al.* (2009) Integration of transport-based models for phyllotaxis and midvein formation. *Genes Dev* **23**: 373–384.

Bergeron F and Reutenauer C. (2019) Golden ratio and phyllotaxis, a clear mathematical link. *J Math Biol* **78**: 1–19.

Bernasconi GP and Boissonade J. (1997) Phyllotactic order induced by symmetry breaking in advected Turing's patterns. *Phys Lett A* **232**: 224–230.

Besnard F, Rafahi Y, Morin V, *et al.* (2014a) Cytokinin signalling inhibitory fields provide robustness to phyllotaxis. *Nature* **505**: 417–421.

Besnard R, Rozier F, and Vernoux T. (2014b) The AHP6 cytokinin signaling inhibitor mediates an auxin-cytokinin cross-talk that regulates the timing of organ initiation at the shoot apical meristem. *Plant Signal Behav* **9**: 4–7.

Beyer R and Richter-Gebert J. (2016) Emergence of complex patterns in a higher-dimensional phyllotactic system. *Acta Soc Bot Pol* **85**(4): 3528.

Bhatia N, Bozorg B, Larsson A, *et al.* (2016) Auxin acts through MONOPTEROS to regulate plant cell polarity and pattern phyllotaxis. *Curr Biol* **26**: 3202–3208.

Bhatia N and Heisler MG. (2018) Self-organizing periodicity in development: organ positioning in plants. *Development* **145**: dev149336. DOI:10.1242/dev.149336.

Bilhuber E, von. (1933) Beiträge zur Kenntnis der Oganstellungen in Pflanzen-reich. *Botanisches Archiv* **35**: 188–250.

Boeyens JCA. (2003) Number patterns in nature. *Crystal Eng* **6**: 167–185.

Bonnet C. (1754) *Recherches sur l'usage des feuilles dans les plantes.* Goettin-gen and Leyden, E. Luzac, fils.

Bowman JL. (1994) *Arabidopsis, an Atlas of Morphology and Development.* Springer-Verlag, New York.

Bowman JL and Esched Y. (2000) Formation and maintenance of the shoot apical meristem. *Trends Plant Sci* **5**: 110–115.

Bradley D, Vincent C, Carpenter R, and Coen E. (1996) Pathways for inflores-cence and floral induction in *Antirrhinum. Development* **122**: 1535–1544.

Braun A. (1831) Vergleichende Untersuchung über die Ordnung der Schuppen an den Tannenzapfen als Einleitung zur Untersuchung der Blattstellung Überhaupt. *Verhandlungen der Kaiserlichen Leopoldinisch-Carolinischen Akademie der Naturforscher* **15**: 195–402.

Braun A. (1835) Dr. Carl Schimper's Vorträge über die Möglichkeit eines wissen-schaftlichen Verstandnis der Blattstellung, nebst Andeutung de Hauptsächlichen Blattstellungsgesetze und Insbesondere der Neuentdeck-ten Gesetze der Aneinanderreihung von Cyclen Verschiedener Maasse. *Flora* **18**: 145–191.

Bravais L and Bravais A. (1837) Essai sur la disposition des feuilles curvisériées. *Annales des Sciences Naturelles Botanique* **7**: 42–110; 193–221; 291–348; **8**: 11–42.

Bravais L and Bravais A. (1839) Essai sur la disposition générale des feuilles rectisériées. *Annales des Sciences Naturelles Botanique* **12**: 5–14; 65–77.

Braybrook SA and Kuhlemeier C. (2010) How a plant builds leaves. *Plant Cell* **22**: 1006–1018.

Braybrook SA and Peaucelle A. (2013) Mechano-chemical aspects of organ formation in *Arabidopsis thaliana*: the relationship between auxin and pectin. *PLOS One* **8**: e57813.

Brillouin L. (1959) *La science et la Théorie de l'Information*. Masson, Paris.

Brooks DR and Wiley EO. (1986) *Evolution as Entropy*. University of Chicago Press, Chicago, London.

Bryntsev VV. (2000) Conformities of spiral phyllotaxis. *Zh Obshch Biol* **61**: 325–335.

Budrene EO and Berg HC. (1991) Complex patterns formed by motile cells of *Escherichia coli*. *Nature* **349**: 630–633.

Bull-Hereñu K and Claßen-Bockhoff R. (2011) Open and closed inflorescences: more than simple opposites. *J Exp Bot* **62**: 79–88.

Burian A, Raczynska-Szajgin M, Borowska-Wykret D, *et al.* (2015) The cup-shaped cotyledon 2 and 3 genes have a post-meristematic effect on *Arabidopsis thaliana* phyllotaxis. *Ann Bot* **115**: 807–820.

Bursill LA and Rouse JL. (1998) Investigation of phyllotaxis of Rhododendron. In: Jean RV, Barabé D (eds), *Symmetry in Plants*. World Scientific, Singapore, pp. 3–32.

Byrne ME, Groover AT, Fontana JR, and Martienssen RA. (2003) Phyllotactic pattern and stem cell fate are determined by the *Arabidopsis* homeobox gene *BELLRINGER*. *Development* **130**: 3941–3950.

Caderas D, Muster M, Vogler H, *et al.* (2000) Limited correlation between expansin gene expression and elongation growth rate. *Plant Physiol* **123**: 1399–1413.

Callos JD, DiRado M, Xu B, *et al.* (1994) The forever young gene encodes an oxidoreductase required for proper development of the *Arabidopsis* vegetative shoot apex. *Plant J* **6**: 835–847.

Callos JD and Medford JI. (1994) Organ positions and pattern formation in the shoot. *Plant J* **6**: 1–7.

Carpenter R, Copsey L, Vincent C, *et al.* (1995) Control of flower development and phyllotaxy by meristem identity genes in *Antirrhinum*. *Plant Cell* **7**: 2001–2011.

Chaitin GJ. (1966) On the length of programs for computing finite binary sequences. *J ACM* **13**: 547–569.

Chaitin GJ. (1987) Information randomness & incompleteness. Papers on algorithmic information theory. *Series in Computer Science,* Vol. 8. World Scientific, Singapore.

Chapman JM and Perry R. (1987) A diffusion model of phyllotaxis. *Ann Bot* **60**: 377–389.

Charlton WA. (1974) Studies in Alismataceae. V. Experimental modification of phyllotaxis in pseudostolons of *Echinodorus tenellus* by means of growth inhibitors. *Can J Bot* **52**: 1131–1142.

Charlton WA. (1978) Studies in Alismataceae. VII. Disruption of phyllotactic and organogenetic patterns in pseudostolons of *Echinoderus tenellus* by mean of growth substances. *Can J Bot* **57**: 215–222.

Charlton WA. (1993) The rotated-lamina syndrome. I. Ulmaceae. *Can J Bot* **71**: 211–221.

Charlton WA. (1998) Pendulum symmetry. In: Jean RV, Barabé D (eds), *Symmetry in Plants*. World Scientific, Singapore, pp. 61–87.

Chaudhury AM, Letham S, Craig S, and Dennis ES. (1993) *Amp1*-A mutant with high cytokinin levels and altered embryonic pattern, faster vegetative growth, constitutive photomorphogenesis and precocious flowering. *Plant J* **4**: 907–916.

Chitwood DH, Headland LR, Rnajan A, *et al.* (2012) Leaf asymmetry as a developmental constraint imposed by auxin-dependent phyllotactic patterning. *Plant Cell* **24**: 2318–2327.

Cho HT and Cosgrove DJ. (2000) Altered expression of expansin modulates leaf growth and pedicel abscission in *Arabidopsis thaliana. Proc Natl Acad Sci* (PNAS) **97**: 9783–9788.

Christenson SK, Dagenais N, Chory J, and Weigel D. (2000) Regulation of the auxin response by the protein kinase PINOID. *Cell* **100**: 469–478.

Chuang CF, Running MP, William-Roberts W, and Meyerowitz EM. (1999) The PERIANTHA gene encodes a bZIP protein involved in the determination of floral organs numbers in *Arabidopsis thaliana. Genes Dev* **13**: 334–344.

Church AH. (1904) *On the Relation of Phyllotaxis to Mechanical Laws*. Williams and Norgate, London.

Church AH. (1920) *On the Interpretation of Phenomena of Phyllotaxis*. Facsimile of the original edition. Hafner Publishing Co., New York, p. 1968.

Clark SE, Jacobsen SE, Levin JZ, and Meyerowitz EM. (1996) The CLAVATA and SHOOT MERISTEMLESS loci competitively regulate meristem activity in *Arabidopsis. Development* **122**: 1567–1575.

Clark SE, Running MP, and Meyerowitz EM. (1993a) CLAVATA1, a regulator of meristem and flower development in *Arabidopsis. Development* **119**: 397–418.

Clark SE, Running MP, and Meyerowitz EM. (1993b) CLAVATA3 is a regulator of shoot and floral meristem development affecting the same processes as CLAVATA1. *Development* **121**: 2057–2067.

Cockcroft CE, den Boer BGW, Healy JMS, and Murray JAH. (2000) Cyclin D controls growth rate in plants. *Nature* **405**: 575–579.

Coen ES, Nugent JM, Luo D, *et al.* (1995) Evolution of floral symmetry. *Philos Trans R Soc London B* **350**: 35–38.

Cooke TJ. (2006) Do Fibonacci numbers reveal the involvement of geometrical imperatives or biological interactions in phyllotaxis? *Bot J Linn Soc* **150**: 3–4.

Couder Y. (1998) Initial transitions, order and disorder in phyllotactic patterns: the ontogeny of *Helianthus annuus*: a case study. *Acta Soc Bot Pol* **67**: 129–150.

Cronk Q and Möller M. (1997) Genetics of floreal symmetry revealed. *TREE* **12**: 85–86.

Cummings FW and Strickland JC. (1998) A model of phyllotaxis. *J Theor Biol* **192**: 531–544.

Dawe RK and Freeling M. (1991) Cell lineage and its consequences in higher plants. *Plant J* **1**: 3–8.

Day SJ and Lawrence PA. (2000) Measuring dimensions: the regulation of size and shape. *Development* **127**: 2977–2987.

De Renzi M. (1997) La forma: sus reglas y evolución, y los datos del registro fósil. In: Aguirre E, Morales J, Soria D (eds), *Registros fósiles e Historia de la Tierra. Cursos de Verano de El Escorial*. Editorial Compluteuse, Madrid, pp. 57–77.

Deb Y, Marti D, Frenz M, *et al.* (2015) Phyllotaxis involves auxin drainage through leaf primordia. *Development* **142**: 1992–2001.

Desmet M and Schaaf A. (1995) Un outil d'analyse spectral des systèmes naturels à pavage régulier: application à la phyllotaxie du *Lepidodendron*. *Comptes Rendus de l'Académie des Sciences*. Paris, Série IIa **320**: 141–144.

Doonan J. (2000) Social controls on cell proliferation in plants. *Curr Opin Plant Biol* **3**: 482–487.

Dormer CJ. (1972) *Shoot Organization in Vascular Plants*. Syracuse University Press, Syracuse.

Dosio GAA, Tardieu F, and Turc O. (2006) How does the meristem of sunflower capitulum cope with tissue expansion and floret initiation? A quantitative analysis. *New Phytol* **170**: 711–722.

Douady S. (1998) The selection of phyllotactic patterns. In: Jean RV, Barabé D (eds), *Symmetry in Plants*. World Scientific, Singapore, pp. 335–358.

Douady S and Couder Y. (1992) Phyllotaxis as a physical self-organized process. *Phys Rev Lett* **68**: 2098–2101.

Douady S and Couder Y. (1996a) Phyllotaxis as a dynamical self-organizing process. Part. I. The spiral modes resulting from time periodic iterations. *J Theor Biol* **178**: 255–274.

Douady S and Couder Y. (1996b) Phyllotaxis as a dynamical self-organizing process. Part. II. The spontaneous formation of a periodicity and the coexistence of spiral and whorled patterns. *J Theor Biol* **178**: 275–294.

Douady S and Couder Y. (1996c) Phyllotaxis as a dynamical self-organizing process. Part. III. The transients leading to the ontogeny. *J Theor Biol* **178**: 295–312.

Douady S and Couder Y. (1998) The phyllotactic pattern as resulting from self-organization in an iterative process. In: Jean RV, Barabé D (eds), *Symmetry in Plants*. World Scientific, Singapore, pp. 539–570.

Douady S and Golé C. (2016) Fibonacci or quasi-symmetric phyllotaxis. Part II: Botanical observations. *Acta Soc Bot Pol* **85**: 3534. DOI:10.5586/asbp.3534.

Doust AN. (2001) The developmental basis of floral variation in *Drymis winteri* (Winteraceae). *Int J Plant Sci* **162**: 697–717.

d'Ovidio F, Anderson CA, Ernstsen CN, and Moselike E. (1999) Bifurcation analysis of spiral growth processes in plants. *Math Comput Simul* **49**: 41–56.

d'Ovidio F and Mosekilde E. (2000) Dynamical system approach to phyllotaxis. *Phys Rev E* **61**: 354–365.

Dumais J and Steele CR. (2000) New evidence for the role of mechanical forces in the shoot apical meristem. *J Plant Growth Regul* **19**: 7–18.

Endress PK. (1989) Chaotic floral phyllotaxis and reduced perianth in *Achlys* (Berberidaceae). *Bot Acta* **102**: 159–163.

Endress PK and Doyle JA. (2007) Floral phyllotaxis in basal angiosperms: development and evolution. *Curr Opin Plant Biol* **10**: 52–57.

Endress PK and Doyle JA. (2009) Reconstructing the ancestral angiosperm flower and its initial specializations. *Amer J Bot* **96**: 22–66.

Erickson RO. (1973) Tubular packing of spheres in biological fine structures. *Science* **181**: 705–716.

Feugier FG, Mochizuki A, and Iwasa Y. (2005) Self-organization of the vascular system in plant leaves: inter-dependant dynamics of auxin flux and carrier proteins. *J Theor Biol* **236**: 366–375.

Fierz V. (2015) Aberrant phyllotactic patterns in cones of some conifers: a quantitative study. *Acta Soc Bot Pol* **84**: 261–265. DOI:105586/asbp.2015.025.

Fisher JB. (1973) Unusual branch development in the palm *Chrysalidocarpus*. *Bot J Linn Soc* **66**: 83–95.

Fleming AJ. (2002) Plant mathematics and Fibonacci flowers. *Nature* **418**: 723.

Fleming AJ, Caderas D, Wehrli E, McQueen MS, *et al.* (1999) Analysis of expansin-induced morphogenesis on the apical meristem of tomato. *Planta* **208**: 166–174.

Fleming AJ, McQueen-Mason S, Mandel T, and Kuhlemeier C. (1997) Induction of leaf primordia by the cell wall protein expansin. *Science* **276**: 1415–1418.

Fleury V. (1999) Un possible lien entre la croissance dendritique en physique et la morphogenèse des plantes. *Comptes Rendus de l'Académie des Sciences*. Paris, Sciences de la vie, **322**: 725–734.

Fujita H and Kawaguchi M. (2018) Spatial regularity control of phyllotaxis pattern generated by the mutual interaction between auxin and PIN1. *PLOS Comput Biol* **14**(4): e1006065. DOI:10.1371/journal.pcbi.1006065.

Fujita H and Mochizuki A. (2006) Pattern formation of leaf veins by the positive feedback regulation between auxin flux and auxin efflux carrier. *J Theor Biol* **241**: 541–551.

Fujita H, Toyokura K, Okada K, and Kawaguchi M. (2011) Reaction–diffusion pattern in shoot apical meristem of plants. *PLOS One* **6**: e18243.

Fujita T. (1942) Zurkenntnis der Orgastellungen im Pflanzenreich. *Jpn J Bot* **12**: 1–55.

Furner IJ and Pumbrey JE. (1992) Cell fate in the shoot apical meristem of *Arabidopsis thaliana*. *Development* **115**: 755–764.

Galvan-Ampudia CS, Chaumeret AM, Godin C, and Vernoux T. (2016) Phyllotaxis: from patterns of organogenesis at the meristem to shoot architecture. *WIREs Dev Biol* **5**: 460–473. DOI:10.1002/wdev.231.

Gatlin LL. (1972) *Information Theory and the Living System*. Columbia University Press, New York.

Giulini A, Wang J, and Jackson D. (2004) Control of phyllotaxy by the cyto-kinin-inducible response regulator homologue ABPHYL1. *Nature* **430**: 1031–1034. DOI:10.1038/nature02778.

Goebel K. (1928) *Organographie der Planzen insbesondere der Archegoni-aten und Samenpflanzen. Teil I. Allgemeine Organographie*, 3rd ed. Gustav Fisher, Jena.

Gola EM and Banasiak A. (2016) Diversity of phyllotaxis in land plants in ref-erence to the shoot apical meristem structure. *Acta Soc Bot Pol* **85**: 3529. DOI:10.5586/asbp.3529.

Golé C, Dumais J, and Douady S. (2016) Fibonacci or quasi-symmetric phyllotaxis. Part 1: Why? *Acta Soc Bot Pol* **85**(4): 3533. DOI:10.5586/asbp.3533.

Goodall C. (1991) Eigenshape analysis of a cut-grow mapping for triangles, and its application to phyllotaxis in plants. *SIAM J Appl Math* **51**: 775–798.

Gordon SP, Chickarmane VS, Ohno C, and Meyerowitz EM. (2009) Multi-ple feedback loops through cytokinin signaling control stem cell number within the Arabidopsis shoot meristem. *Proc Natl Acad Sci (PNAS)* **106**: 16529–16534.

Green PB. (1987) Inheritance of pattern: analysis from phenotype to gene. *Am Zool* **27**: 657–673.

Green PB. (1992a) Pattern formation in shoots: a likely role for minimal energy configurations of the tunica. *Int J Plant Sci* **153**: S59–S75.

Green PB. (1992b) Expressions of form and pattern in plants: a role for biophysical fields. *Cell Dev Biol* **7**: 903–911.

Green PB. (1996) Transductions to generate plant form and pattern: an essay on cause and effect. *Ann Bot* **78**: 269–281.

Green PB. (1999) Expression of patterns in plants: combining molecular and calculus-based biophysical paradigms. *Am J Bot* **86**: 1059–1076.

Green PB and Baxter DR. (1987) Phyllotactic patterns: characterization by geo-metrical activity at the formative region. *J Theor Biol* **128**: 387–395.

Green PB and Linstead P. (1990) A procedure for SEM of complex shoot structures applied to the inflorescence of snapdragon (*Antirrhinum*). *Protoplasma* **158**: 33–38.

Green PB, Steele R, and Rennich SC. (1998) How plants produce patterns. A review and a proposal that undulating field is the mechanism. In: Jean

RV, Barabé D (eds), *Symmetry in Plants*. World Scientific, Singapore, pp. 359–392.

Greyson RI and Walden DB. (1972) The ABPHYL syndrome in *Zea mays*. I. Arrangement, number and size of leaves. *Amer J Bot* **59**: 466–472.

Guédon Y, Refahi Y, Besnard F, *et al*. (2013) Pattern identification and characterization reveal permutations of organs as a key genetically controlled property of post-meristematic phyllotaxis. *J Theor Biol* **338**: 94–110.

Guenot B, Bayer E, Kierzkowski D, *et al*. (2014) PIN1-independent leaf initiation in *Arabidopsis*. *Plant Physiol* **159**: 1501–1510.

Guerreiro J. (1995) Phyllotaxis: an interdisciplinary phenomenon. *Physica D* **80**: 356–384.

Guerreiro J and Rothen F. (1998) Universal results from a simple model of phyllotaxis. In: Jean RV, Barabé D (eds), *Symmetry in Plants*. World Scientific, Singapore, pp. 571–600.

Hamant O. (2017) Mechano-devo. *Mech Dev* **145**: 2–9.

Hamant O, Heisler MG, Jönsson H, *et al*. (2008) Developmental patterning by mechanical signals in *Arabidopsis*. *Science* **322**: 1650–1655. DOI:10.1126/science.1165594.

Hardtke CS and Berleth T. (1998) The *Arabidopsis* gene, MONOPTEROS, encodes a transcription factor mediating embryo axis formation and vascular development. *EMBO J* **17**: 1405–1411.

Hauk J and Mika K. (2003) Attractive or repulsive interactions in different structure types. *Prog Solid State Ch* **31**: 149–218.

Heisler M, Hamant O, Krupinski P, *et al*. (2010) Alignment between PIN1 polarity and microtubule orientation in the shoot apical meristem reveals a tight coupling between morphogenesis and auxin transport. *PLOS Biol* **8**(10): e1000516. DOI: 10.1371/journalpbio.1000516.

Heisler MG and Jönsson H. (2006) Modeling auxin transport and plant development. *J Plant Growth Regul* **25**: 302–312.

Heisler MG, Ohno C, Das P, *et al*. (2005) Patterns of auxin transport and gene expression during primordium development revealed by live imaging of the Arabidopsis inflorescence meristem. *Curr Biol* **15**: 1899–1911.

Hejnowicz Z and Sievers A. (1996) Tissue stresses in organs of herbaceous plants, III: elastic properties of the tissues of the sunflower capitulum. *J Exp Biol* **47**: 519–528.

Hellendoorn PH and Lindenmayer A. (1974) Phyllotaxis in *Bryophyllum tubiflorum*: morphogenetic studies and computer simulations. *Acta Botanica Neerlandica* **23**: 473–492.

Hellwig H, Engelmann R, and Deussen O. (2006) Contact pressure models for spiral phyllotaxis and their computer simulation. *J Theor Biol* **240**: 489–500.

Hellwig H and Neukirchner T. (2010) Phyllotaxis. Die mathematische Beschreibung und Modellierung von Blattstellungsmustern. *Math Semesterber* **57**: 17–56.

Hernandez LH and Green PB. (1993) Transductions for the expression of structural pattern: analysis in sunflower. *Plant Cell* **5**: 1725–1738.

Hirmer M. (1922) *Zur Lösung des Problems der Blattstellungen*. Jena, Gustav Fisher, p. 109.

Hofmeister W. (1868) Allgemeine Morphologie der Gewachse. In: Bary Ade, Irmisch TH, Sachs J (eds), *Handbuch der Physiologischen Botanik*. Engelmann, Leipzig, pp. 405–664.

Hotton S. (1999) *Symmetry of Plants*. University of California, Santa Cruz. Thesis.

Hotton S. (2003) Finding the center of phyllotaxis pattern. *J Theor Biol* **225**: 15–32.

Hotton S. (2008) Applying the dynamics of circle maps to phyllotactic patterns. http://maven.smith.edu/~phyllo/Assets/elliptalk/talk.i.html.

Hotton S, Johnson V, Wilbarger J, *et al.* (2006) The possible and the actual in phyllotaxis: bridging the gap between empirical observations and iterative methods. *J Plant Growth Regul* **25**: 313–323. DOI:10.1007/s00344-006-0067-9.

Itoh J, Hibara K, Kojima M, *et al.* (2012) Rice DECUSSATE controls phyllotaxy by affecting the cytokinin signaling pathway. *Plant J* **72**: 869–881.

Itoh JI, Hasegawa A, Kitano H, and Nagato Y. (1998) A recessive heterochronic mutation, *plastochron1*, shortens the plastochron and elongates the vegetative phase in rice. *Plant Cell* **10**: 1511–1521.

Itoh JI, Kitano H, Matsuoka M, and Nagato Y. (2000) *SHOOT ORGANIZATION* Genes regulate shoot apical meristem organization and the pattern of leaf primordium initiation in rice. *The Plant Cell* **12**: 2161–2174.

Jackson D and Hake S. (1999) Control of phyllotaxis in maize by the *abphyl1* gene. *Development* **126**: 315–323.

Jackson M, Duran-Nebreda S, Kierzkowski D, *et al.* (2019) Global topological order emerges through local mechanical control of cell divisions in the *Arabidopsis* shoot apical meristem. *Cell Syst* **8**: 53–65.

Jean RV. (1976) La G-entropie en phyllotaxie. *Rev BioMath* **55**: 111–130.

Jean RV. (1979) Some consequences of the hierarchical approach to phyllotaxis. *J Theor Biol* **81**: 309–326.

Jean RV. (1980) A systemic model of growth in botanometry. *J Theor Biol* **87**: 569–584.

Jean RV. (1986) An interpretation of Fujita's frequency diagrams frequency diagrams in phyllotaxis. *Bull Math Biol* **48**: 77–86.

Jean RV. (1988) Number-theoretic properties of two dimensional lattices. *J Number Theory* **29**: 206–223.

Jean RV. (1994) *Phyllotaxis: A Systemic Study in Plant Morphogenesis.* Cambridge University Press, Cambridge, New York.

Jean RV. (1998a) Elementary rules of growth in phyllotaxis. In: Jean RV, Barabé D (eds), *Symmetry in Plants.* World Scientific, Singapore pp. 601–617.

Jean RV. (1998b) Cross-fertilization between models in phyllotaxis. *J Biol Syst* **7**: 145–158.

Jean RV and Barabé D. (1998b) Phyllotaxis — the way ahead, a view on open questions and directions of research. *J Biol Syst* **6**: 95–126.

Jean RV and Barabé D. (2001) Application of two mathematical models to a family of plants with enigmatic phyllotactic processes. *Ann Bot* **88**: 173–186.

Jean RV and Barabé D (eds). (1998a) *Symmetry in Plants.* World Scientific, Singapore.

Jeune B and Barabé D. (2004) Statistical recognition of random and regular phyllotactic patterns. *Ann Bot* **94**: 913–917.

Jeune B and Barabé D. (2006a) A stochastic approach to phyllotactic patterns analysis. *J Theor Biol* **238**: 52–59.

Jeune B and Barabé D. (2006b) Simulations of transitions from regular to stochastic phyllotactic patterns. *J Biol Syst* **14**: 113–129.

Jones CS and Drinnan AN. (2009) The developmental pattern of shoot apices in *Selaginella kraussiana* (Kunze) A. Braun. *Int J Plant Sci* **170**: 1009–1018.

Jönsson H, Heisler MG, Shapiro BE, *et al.* (2006) An auxin-driven polarized transport model for phyllotaxis. *Proc Natl Acad Sci (PNAS)* **103**: 1633–1638.

Kaplan DR. (1992) The relationship of cells to organisms in plants: problem and implications of an organismal perspective. *Int J Plant Sci* **153**: S28–S37.

Kaplan DR, Dengler NG, and Dengler RE. (1982) The mechanism of placation inception in Palm leaves: problem and developmental morphology. *Can J Bot* **60**: 2939–2975.

Kappraff J, Blackmore D, and Adamson G. (1998) Phyllotaxis as a dynamic system: a study in number. In: Jean RV, Barabé D (eds), *Symmetry in Plants*. World Scientific, Singapore, pp. 409–458.

Karmakar S and Key ES. (2004) Compositions of random Möbius transformations. *Stoch Anal Appl* **22**: 525–557.

Kaya H, Shibahara KI, Taoka KI, *et al.* (2001) *Fasciata* genes for chromatin assembly factor-1 in *Arabidopsis* maintain the cellular organization of apical meristems. *Cell* **104**: 131–142.

Kelly A, Bonnlander MB, and Meeks-Wagner DR. (1995) NFL, the tobacco homolog of Floricaula and Leafy, is transcriptionally expressed in both vegetative and floral meristems. *Plant Cell* **7**: 225–234.

Kelly WJ and Cooke TJ. (2003) Geometrical relationships specifying the phyllotactic patterns of aquatic plants. *Amer J Bot* **90**: 1131–1143.

Kierzkowski D, Lenhard M, Smith R, and Kuhlemeier C. (2013) Interaction between meristem and tissue layers controls phyllotaxis. *Developmental Cell* **26**: 616–628.

Kierzkowski D, Nakayama N, Routier-Kierzkowska AL, *et al.* (2012) Elastic domains regulate growth and organogenesis in the plant shoot apical meristem. *Science* **335**: 1096–1099.

Kierzkowski D and Routier-Kierzkowska AL. (2019) Cellular basis of growth in plants: geometry matters. *Curr Opin Plant Biol* **47**: 56–63.

King S, Beck F, and Lüttge U. (2004) On the mystery of the golden angle in phyllotaxis. *Plant Cell Environ* **27**: 685–695.

Kirchoff BK. (2000) Hofmeister's rule and primordium shape: constraints on organ position in *Hedychium coronarium* (Zingiberaceae). In: Wilson KL, Morrison DA (eds), *Monocots: Systematics and Evolution*. CSIRO, Collingwood, pp. 75–83.

Kirchoff BK. (2003) Shape matter: Hofmeister's rule, primordium shape, and flower orientation. *Int J Plant Sci* **164**: 505–517.

Kitazawa MS and Fujimoto K. (2015) A dynamical phyllotaxis model to determine floral organ number. *PLOS Comput Biol.* DOI:10.13711/journal. pcbi.1004145.

Kitazawa MS and Fujimoto K. (2018) Spiral phyllotaxis underlies constrained variation in *Anemone* (Ranunculaceae) tepals arrangement. *J Plant Res* **131**: 459–468.

Klar AJS. (2002) Fibonacci's flowers. *Nature* **417**: 595.

Koch AJ, Bernasconi G, and Rothen F. (1998) Phyllotaxis as a geometrical and dynamical system. In: Jean RV, Barabé D (eds), *Symmetry in Plants*. World Scientific, Singapore, pp. 459–486.

Kolmogorov AN. (1965) Three approaches to the quantitative definition of information. *Probl Inform Transm* **1**: 1–7.

Korn RW. (2006) Anodic asymmetry of leaves and flowers and its relationship to phyllotaxis. *Ann Bot* **97**: 1011–1015.

Korn RW. (2008) Phyllotaxis: theories and evolution. *Int J Plant Dev Biol* **2**: 1–12.

Kramer EM. (2004) PIN and AUX/LAX proteins: their role in auxin accumulation. *Trends Plant Sci* **9**: 578–582.

Kramer EM. (2008) Computer models of auxin transport: a review and commentary. *J Exp Bot* **59**: 45–53.

Kuhlemeier C. (2007) Phyllotaxis. *Trends Plant Sci* **12**: 143–150.

Kunz M. (1995) Some analytic results about two physical models of phyllotaxis. *Comm Math Phys* **169**: 261–295.

Kunz M. (1997) Phyllotaxie, billards polygonaux et théorie des nombres. Université de Lausanne. Thèse.

Kunz M. (2001) Dynamical models of phyllotaxis. *Physica D* **157**: 147–165.

Kwiatkowska D. (2004) Structural integration at the shoot apical meristem: models, measurements, and experiments. *Amer J Bot* **91**: 1277–1293.

Kwiatkowska D and Florek-Marwitz J. (1999) Ontogenetic variation of phyllotaxis and apex geometry in vegetative shoots of *Sedum maximum* (L.) Hoffm. *Acta Soc Bot Pol* **68**(2): 85–95.

Lacroix C, Barabé D, and Jeune B. (2004) Early stages of initiation of two types of leaves in *Thuja occidentalis* (eastern white cedar). *Can J Bot* **82**: 598–606.

Lacroix C, Jeune B, and Barabé D. (2005) Encasement in plant morphology: an integrative approach from genes to organisms. *Can J Bot* **83**: 1207–1221.

Lacroix CR, Steeves R, and Kemp JF. (2007) Floral development, fruit set, and dispersal of the Gulf of St. Lawrence Aster (*Symphyotrichum laurentianum*) (Fernald) Nesom. *Can J Bot*. **85**: 331–341. DOI:10.1139/B07-011.

Landrein B and Hamant O. (2013) How mechanical stress controls microtubule behavior and morphogenesis in plants: history, experiments and revisited theories. *Plant J* **75**: 324–338.

Landrein B, Refahi Y, Besnard F, *et al.* (2015) Meristem size contributes to the robustness of phyllotaxis in *Arabidopsis*. *J Exp Bot* **66**: 1317–1324. DOI:10.1093/jxb/eru482.

Laufs P, Grandjean O, Jonak C, *et al.* (1998) Cellular parameters of the shoot apical meristem in *Arabidopsis*. *The Plant Cell* **10**: 1375–1389.

Laux T, Mayer KFX, Berger J, and Jurgens G. (1996) The WUSCHEL gene is required for shoot and floral meristem identity in *Arabidopsis*. *Development* **122**: 87–96.

Le Guyader H. (1988) La phyllotaxie ou le rêve du crystal vivant. In: Barreau H (ed), *Théories Biologiques — Éthique et Expérimentation en Médecine*. Éditions du CNRS, Paris, pp. 111–153.

Lee HW and Levitov LS (1998) Universality in phyllotaxis: a mechanical theory. In: Jean RV, Barabé D (eds), *Symmetry in Plants*. World Scientific, Singapore, pp. 619–653.

Lestiboudois MT. (1848) Phyllotaxie anatomique. *Ann Sci Nat* **3**: 15–105.

Levitov LS. (1991a) Phyllotaxis of flux lattices in layered superconductors. *Phys Rev Lett* **66**: 224–227.

Levitov LS. (1991b) Energetic approach to phyllotaxis. *Europhys Lett* **14**: 533–539.

Leyser HMO and Fumer IJ. (1992) Characterization of three apical meristem mutants of *Arabidopsis thaliana*. *Development* **122**: 87–96.

Li C, Zhang X, and Cao Z. (2005) Triangular and Fibonacci patterns driven by stress on core/shell microstructures. *Science* **309**: 909–911.

Liaw SS. (1998) Phyllotaxis: its geometry and dynamics. *Phys Rev E* **57**: 4589–4593.

Lieber MM. (1998) The living spiral — a dimensionless biological constant gives a new perspective to physics. *Riv Biol Forum* **91**: 91–118.

Linardic M and Braybrook SA. (2017) Towards an understanding of spiral patterning in the *Sargassum muticum* shoot apex. *Sci Rep* **7**: 13887–13896. DOI:10.1038/s41598-017-13767-5.

Liu CM, Zhi-Hong Xu ZH, and Chua NH. (1993) Auxin polar transport is essential for the establishment of bilateral symmetry during early plant embryogenesis. *Plant Cell* **5**: 621–630.

Liu RT and Liaw SY. (2007) Whorled patterns of phyllotaxis. *Int J Mod Phys C* **18**: 1811–1817.

Loiseau JE. (1969) *La phyllotaxie*. Masson et Cie, Paris.

Long J and Barton MK. (2000) Initiation of axillary and floral meristems in *Arabidopsis*. *Dev Biol* **218**: 341–353.

Luo D, Carpenter R, Vincent C, *et al.* (1996) Origin of floral symmetry in *Antirrhinum*. *Nature* **383**: 794–799.

Lyndon RF. (1998) Phyllotaxis in flowers and in flower reversion. In: Jean RV, Barabé D (eds), *Symmetry in Plants*. World Scientific, Singapore, pp. 109–124.

Macia E, Domínguez-Adame F, and Sánchez A. (1994) Energy spectra of quasiperiodic systems via information entropy. *Phys Rev E* **50**: 679–682.

Mähönen AP, Higuchi M, Törmäkangas K, *et al.* (2006) Cytokinins regulate a bidirectional phosphorelay network in *Arabidopsis*. *Curr Biol* **16**: 1116–1122. DOI 10.1016/j.cub.2006.04.030.

Maksymowych R and Erickson R. (1977) Phyllotactic change induced by gibberellic acid in *Xanthium*. *Amer J Bot* **64**: 33–44.

Malygin AG. (1998) Structural theory of phyllotaxis I. The mechanism of formation of the spiral structures of alternate phyllotaxis. *Biofizika* **43**: 335–342.

Malygin AG. (2000) Structural theory in phyllotaxis. II. Interrelation between lower and superior phyllotaxis forms. *Biofizika* **45**: 1112–1118.

Marc J and Hackett WP. (1991) Gibberellin-induced reorganization of spatial relationships of emerging leaf primordia at the shoot apical meristem in *Hedera helix* L. *Planta* **185**: 171–178.

Marzec C. (1987) Phyllotaxis as a dissipative structure. *Math Mod* **8**: 740–745.

Marzec C. (1998) Uniform spacing models for the morphogenesis of high symmetry biological structures: icosahedral capsids, coaxial helices, and helical phyllotaxis. In: Jean RV, Barabé D (eds), *Symmetry in Plants*. World Scientific, Singapore, pp. 681–722.

Marzec C. (1999a) A pragmatic approach to modelling morphogenesis. *J Biol Syst* **7**: 333–351.

Marzec C. (1999b) The morphogenesis of high symmetry: the warping theorem. *J Biol Syst* **7**: 353–427.

Marzec C and Kappraff J. (1983) Properties of maximal spacing on a circle related to phyllotaxis and to the golden mean. *J Theor Biol* **103**: 201–226.

Matkowski A, Karwowski R, and Zagórska-Marek B. (1998) Two algorithms of determining the middle point of the shoot apex by surrounding organ primordia positions and their usage for computer measurements of divergence angles. *Acta Soc Bot Pol* **67**: 151–159.

Mauseth JD. (2004) Giant shoot apical meristems in cacti have ordinary leaf primordia but altered phyllotaxy and shoot diameter. *Ann Bot* **94**: 145–153.

Medina ML. (2010) Two 'EvoDevos'. *Biol Theor* **5**: 7–11.

Meicenheimer R. (1981) Changes in *Epilobium* phyllotaxy induced by N-1-naphthylphalamic acid and α-4-chlorophenoxyisobutyric acid. *Amer J Bot* **68**: 1139–1154.

Meicenheimer R. (1987) Role of stem growth in *Linum usitatissimum* leaf trace patterns. *Amer J Bot* **74**: 857–867.

Meicenheimer R. (1998) Decussate to spiral transitions in phyllotaxis. In: Jean RV, Barabé D (eds), *Symmetry in Plants*. Singapore: World Scientific pp. 125–143.

Meicenheimer R and Zagórska-Marek B. (1989) Consideration of the geometry of the phyllotaxic pattern triangular unit and discontinuous phyllotactic transitions. *J Theor Biol* **139**: 359–368.

Meinhardt H. (1982) *Models of Biological Pattern Formation*. Academic Press, London, UK.

Meinhardt H. (2003) Complex pattern formation by a self-destabilization of established patterns: chemotactic orientation and phyllotaxis as examples. *C R Biol* **326**: 223–237.

Meinhardt H. (2004) Out-of-phase oscillations and traveling waves with unusual properties: the use of three-component systems in biology. *Physica D* **199**: 264–277.

Meinhardt H, Koch AJ, and Bernasconi GP. (1998) Models of pattern formation applied to plant development. In: Jean RV, Barabé D (eds), *Symmetry in Plants*. World Scientific, Singapore, pp. 723–758.

Merks RMH, Van de Peer Y, Inzé D, and Beemster GTS. (2007) Canalization without flux sensors: a travelling-wave hypothesis. *Trends Plant Sci* **12**: 390.

Milani P, Gholamirad M, Traas J, *et al.* (2011) In vivo analysis of local wall stiffness at the shoot apical meristem in *Arabidopsis* using atomic force microscopy. *Plant J* **67**: 1116–1123.

Milani P, Mirabet V, Cellier C, *et al.* (2014) Matching patterns of gene expression to mechanical stiffness at cell resolution through quantitative tandem epifluorescence and nanoindentation. *Plant Physiol* **165**: 1399–1408.

Mirabet V, Besnard F, Vernoux T, and Boudaoud A. (2012) Noise and robustness in phyllotaxis. *PLOS Comput Biol* **8**: e1002389. DOI:10.1371/journal.pcbi.1002389.

Miri M and Rivier N. (2002) Continuum elasticity with topological defects, including dislocations and extra-matter. *J Phys A: Math Theor* **35**: 1727–1739.

Mitchison GJ. (1977) Phyllotaxis and the Fibonacci series. *Science* **196**: 270–275.

Mitchison GJ. (1980) Model for vein formation in higher plants. *Proc R Soc London B, Biol Sci* **207**: 79–109.

Mitchison GJ. (1981) The polar transport of auxin and vein patterns in plants. *Philos Trans R Soc London B, Biol Sci* **295**: 461–471.

Mughal A and Weaire D. (2017) Phyllotaxis, disk packing, and Fibonacci numbers. *Phys Rev E* **95**: 022401.

Nagasawa N, Mioshi M, Kitano H, *et al.* (1996) Mutations associated with floral organ number in rice. *Planta* **198**: 627–633.

Nakayama N and Kuhlemeier C. (2008) Leaf development: untangling the spirals. *Curr Biol* **19**: R71–R74.

Nakayama N, Smith RS, Mandel T, *et al.* (2012) Mechanical regulation of auxin-mediated growth. *Curr Biol* **22**: 1468–1476.

Newell AC and Pennybacker M. (2013) Fibonacci patterns: common or rare? *Procedia IUTAM* **9**: 86–109.

Newell AC and Shipman PD. (2005) Plants and Fibonacci. *J Stat Phys* **121**: 937–968.

Newell AC and Shipman PD. (2008) A new invariant in plant phyllotaxis. *Anal Appl* **6**: 383–399.

Newell AC, Shipman PD, and Cooke TJ. (2012) Patterns on desert plants. *Desert Plants* **28**: 7–21.

Newell AC, Shipman PD, and Sun Z. (2008a) Phyllotaxis: cooperation and competition between mechanical and biochemical processes. *J Theor Biol* **251**: 421–439.

Newell AC, Shipman PD, and Sun Z. (2008b) Article addendum. Phyllotaxis as an example of the symbiosis of mechanical forces and biochemical processes in living tissue. *Plant Signal Behav* **3**: 586–589.

Niklas KJ. (1998) Light harvesting 'fitness landscape' for vertical shoots with different phyllotactic systems. In: Jean RV, Barabé D (eds), *Symmetry in Plants*. World Scientific, Singapore, pp. 759–773.

Nishimura A, Ito M, Kamiya N, *et al.* (2002) *OsPNH1* regulates leaf development and maintenance of the shoot apical meristem in rice. *Plant J* **30**: 189–201.

Nisoli C, Gabor NM, Lammert PE, *et al.* (2009) Static and dynamical phyllotaxis in a magnetic cactus. *Phys Rev Lett* **102**: 186103.

Nisoli C, Gabor NM, Lammert PE, *et al.* (2010) Annealing a magnetic cactus. *Phys Rev E* **81**: 046107.

Obara M, Ikeda K, Itoh JI, and Nagato Y. (2004) Characterization of leaf lateral symmetry 1 mutant in rice. *Breeding Sci* **54**: 157–163.

Okabe T, Ishida A, and Yoshimura J. (2019) The unified rule of phyllotaxis explaining both spiral and non-spiral arrangements. *J Royal Soc Interface* **16**: 20180850.

Okabe T. (2011) Physical phenomenology of phyllotaxis. *J Theor Biol* **280**: 63–75.

Okabe T. (2012) Systematic variations in divergence angle. *J Theor Biol* **313**: 20–41.

Okabe T. (2015a) Biophysical optimality of the golden angle in phyllotaxis. *Sci Rep* **5**: 15358. DOI:10.1038/srep15358.

Okabe T. (2015b) Extraordinary accuracy in floret position of *Helianthus annuus*. *Acta Soc Bot Pol* **84**: 79–85.

Okabe T. (2016) The riddle of phyllotaxis: exquisite control of divergence angle. *Acta Soc Bot Pol* **85**: 3527. DOI:10.5586/asbp3527.

Otsuga D, DeGuzman B, Prigge MJ, *et al.* (2001) REVOLUTA regulates meristem initiation at lateral positions. *Plant J* **25**: 223–236.

Palauqui JC and Laufs P. (2011) Phyllotaxis: in search of the golden angle. *Curr Biol* **21**: 502–504.

Paponov IA, Teale WD, Trebar M, *et al.* (2005) The PIN auxin efflux facilitators: evolutionary and functional perspectives. *Trends Plant Sci* **10**: 170–177. DOI:10.106/j.tplants.2005.02.009.

Pautler M, Eveland AL, LaRue T, *et al.* (2015) *FASCIATED EAR4* encodes a bZIP transcription factor that regulates shoot meristem size in maize. *Plant Cell* **27**: 104–120.

Peaucelle A and Couder Y. (2016) Fibonacci spirals in a brown alga [*Sargassum muticum* (Yendo) Fensholt] and in land plant [*Arabidopsis thaliana* (L.) Heynh.]: a case of morphogenetic convergence. *Acta Soc Bot Pol* **85**: 3526. DOI:10.5586/asbp.3526.

Peaucelle A, Louvet R, Johansen JN, *et al.* (2011) The transcription factor BELLRINGER modulates phyllotaxis by regulating the expression of a pectin methylesterase in *Arabidopsis*. *Development* **138**: 4733–4741.

Peaucelle A, Louvet R, Jorunn J, *et al.* (2008) Phyllotaxis in *Arabidopsis* is controlled by the methyl-esterification status of cell wall pectins. *Curr Biol* **18**: 1943–1948.

Peaucelle A, Morin H, Traas J, and Laufs P. (2007) Plants expressing a miR164-resistant CUC2 gene reveal the importance of post-meristematic maintenance of phyllotaxy in *Arabidopsis*. *Development* **134**: 1045–1050.

Pennybacker M and Newell AC. (2013) Phyllotaxis, pushed pattern-forming fronts, and optimal packing. *Phys Rev Lett* **110**: 248104.

Pennybacker MF, Shipman PD, and Newell AC. (2015) Phyllotaxis: some progress, but a story far from over. *Phys D* **306**: 48–81.

Péret B, Li G, Zhao J, *et al.* (2012) Auxin regulates aquaporin function to facilitate lateral root emergence. *Nat Cell Biol* **14**: 991–998.

Petrov EA. (2001) Genesis of morphometric invariants of higher plants as a database for the genesis of morphometric invariants of vertebrates. *Russ J Plant Physiol* **48**: 825–827.

Pien S, Wyrzykowska J, McQueen-Mason S, *et al.* (2001) Local expression of expansin induces the entire process of leaf development and modifies leaf shape. *Proc Natl Acad Sci (PNAS)* **98**: 11812–11817.

Pinon V, Prasad K, Grigg SP, *et al.* (2013) Local auxin biosynthesis regulation by PLETHORA transcription factors controls phyllotaxis in *Arabidopsis*. *Proc Natl Acad Sci (PNAS)* **110**: 1107–1112.

Plantefol L. (1948) Fondements d'une théorie phyllotaxique nouvelle: La théorie des helices foliaires multiples, Masson et Cie, Paris.

Poething RS. (1987) Clonal analysis of cell lineage patterns in plant development. *Amer J Bot* **74**: 581–594.

Prasad K, Grigg SP, Barkoulas M, *et al.* (2011) *Arabidopsis* PLETHORA transcription factors control phyllotaxis. *Curr Biol* **21**: 1123–1128.

Priestly JH. (1928) The meristematic tissues of the plant. *Ann Bot* **3**: 1–20.

Priestly JH and Scott LI. (1933) Phyllotaxis in the dicotyledon from the standpoint of developmental anatomy. *Biol Rev* **8**: 241–268.

Prusinkiewicz P and Lindenmayer A. (1990) *The Algorithmic Beauty of Plants.* Springer-Verlag, New York.

Prusinkiewicz P and Runions A. (2012) Computational models of plant development and form. *New Phytol* **193**: 549–569. DOI:10.1111/j.1469-8133.2011.04009.x.

Ragni L, Belles-Boix E, Günl M, and Pautot V. (2008) Interaction of KNAT6 and KNAT2 with BREVIPEDICELLUS and PENNYWISE in *Arabidopsis* Inflorescences. *Plant Cell* **20**: 888–900. DOI:10.1105/tpc.108.058230.

Refahi Y, Brunoud G, Farcott E, *et al.* (2016) A stochastic multicellular model identifies biological watermarks from disorders in self-organized patterns of phyllotaxis. *eLife* **5**: e14093. DOI:10.7554/eLife.14093.

Refahi Y, Farcot E, Guédon Y, *et al.* (2011) A combinatorial model of phyllotaxis perturbations in *Arabidopsis thaliana.* In: Giancarlo R, Manzini G (eds), *Combinatorial Pattern Matching (CPM) 2011. Lecture Notes in Computer Science 6661.* Springer, Berlin, Heidelberg, pp. 323–335.

Refahi Y, Guédon Y, Besnard F, *et al.* (2010) Analyzing perturbations in phyllotaxis of Arabidopsis thaliana. In: DeJong T, Da Silva D (eds), *6th International Workshop on Functional-Structural Plant Models*, Davis, CA, pp. 185–187.

Reick CH. (2002) Two notions of conspicuity and the classifications of phyllotaxis. *J Theor Biol* **215**: 263–271.

Reick, CH. (2015) A renormalization approach to the universality of scaling in phyllotaxis. *PhysicaD: Nonlinear Phenomena* **298**: 68–86.

Reinhardt D. (2005a) Regulation of phyllotaxis. *Int. J. Dev Biol* **49**(5–6): 539–546.

Reinhardt D. (2005b) Phyllotaxis — a new chapter in an old tale about beauty and magic numbers. *Curr Opin Plant Biol* **8**: 487–493.

Reinhardt D, Frenz M, Mandel T, and Kuhlemeier C. (2005) Microsurgical and laser ablation analysis of leaf positioning and dorsoventral patterning in tomato. *Development* **132**(1): 15–26.

Reinhardt D and Kuhlemeier C. (2002) Phyllotaxis in higher plants. In: McManus MT, Veit BE (eds), *Meristematic Tissues in Plant Growth and Development*. Sheffield Academic Press Ltd., Sheffield, UK, pp. 172–212.

Reinhardt D, Kuhlemeier C, and Smith RS. (2012) Elastic domains regulate growth and organogenesis in the plant shoot apical meristem. *Science* **335**: 1096–1099.

Reinhardt D, Mandel T, and Kuhlemeier C. (2000) Auxin regulates the initiation and radial position of plant lateral organs. *Plant Cell* **12**: 507–518.

Reinhardt D, Pesce ER, Stieger P, *et al.* (2003) Regulation of phyllotaxis by polar auxin transport. *Nature* **426**: 255–260.

Reinhardt D, Witter F, Mandel T, and Kuhlemeier C. (1998) Localized upregulation of a new expansin gene predicts the site of leaf formation in the tomato meristem. *Plant Cell* **10**: 1427–1437.

Reyes E, Nadot S, von Balthazar M, *et al.* (2018) Testing the impact of morphological rate heterogeneity on ancestral state reconstruction of five floral traits in angiosperms. *Sci Rep* **8**: 9473.

Ricard J. (2003) What do we mean by biological complexity? *C R Biol* **326**: 133–140.

Richards FJ. (1948) The geometry of phyllotaxis and its origin. *Symp Soc Exp Biol* **2**: 217–245.

Richards FJ. (1951) Phyllotaxis: its quantitative expression and relation to growth in the apex. *Phil Trans R Soc London* **235B**: 509–564.

Richter PH and Schranner R. (1978) Leaf arrangement, geometry, morphogenesis, and classification. *Naturwissenschaften* **65**: 319–327.

Ridley JN. (1982) Computer simulation of contact pressure in capitula. *J Theor Biol* **95**: 1–11.

Rivier N, Sadoc JF, and Charvolin J. (2016) Phyllotaxis: a framework for foam topological evolution. *Eur Phys J E* **39**: 7. DOI:10.1140/epje/i2016-16007-8.

Robinson S, Burian A, Couturier E, *et al.* (2013) Mechanical control of morphogenesis at the shoot apex. *J Exp Bot* **64**: 4729–4744.

Robinson S, Huflejt M, Barbier de Reuille P, *et al.* (2017) An automated confocal micro-extensometer enables *in vivo* quantification of mechanical properties with cellular resolution. *Plant Cell* **29**: 2959–2973.

Rolland-Lagan AG and Prusinkiewicz P. (2005) Reviewing models of auxin canalization in the context of leaf vein formation in *Arabidopsis*. *Plant J* **44**: 854–865.

Routier-Kierzkowska AL and Runions A. (2018) Modeling plant morphogenesis: an introduction. In: Geitmann A, Gril J (eds), *Cham, Switzerland: Plant Biomechanics*, Springer, pp. 165–192.

Rueda-Contreras MD and Aragón JL. (2014) Alan Turing's chemical theory of phyllotaxis. *Rev Mex Fis* **60**: 1–12.

Rueda-Contreras MD, Romero-Arias JR, Aragón JL, and Barrio RA. (2018) Curvature-driven spatial patterns in growing 3D domains: a mechanochemical model for phyllotaxis. *PLOS One* **13**: e0201746. DOI:10.1371/journal.pone.0201746.

Running MP and Meyerowitz EM. (1996) Mutations in *PERIANTHIA* gene of *Arabidopsis* specifically alter floral organ number and initiation pattern. *Development* **122**: 1261–1269.

Rutishauser R. (1981) Blattstellung und Sprossentwicklung bei Blütenpflanzen. *Dissertationes Botanicae* **62**. Cramer, Vaduz.

Rutishauser R. (1982) Der Plastochrone quotient als Teil einer quantitativen Blattstellungsanalyse bei Samenpflanzen. *Beitr Biol Pflanz* **57**: 323–57.

Rutishauser R. (1998) Plastochrone ratio and leaf arc as parameters of a quantitative phyllotaxis analysis in vascular plants. In: Jean RV, Barabé D (eds), *Symmetry in Plants*. World Scientific, Singapore, pp. 171–212.

Rutishauser R. (2016) Acacia (wattle) and Cananga (ylang-ylang): from spiral to whorled and irregular (chaotic) phyllotactic patterns — a pictorial report. *Acta Soc Bot Pol* **85**: 3531. DOI:10.5586/asbp.3531.

Rutishauser R and Sattler R. (1985) Complementarity and heuristic value of contrasting models in structural botany. I. General considerations. *Bot Jahrb Syst* **107**: 415–455.

Sachs T. (1969) Polarity and the induction of organized vascular tissues. *Ann Bot* **33**: 263–275.

Sachs T. (1981) The control of the patterned differentiation of vascular tissues. *Adv Bot Res* **9**: 151–262.

Sahlin P, Söderberg B, and Jönsson H. (2009) Regulated transport as a mechanism for pattern generation: capabilities for phyllotaxis and beyond. *J Theor Biol* **258**: 60–70.

Sassi M, Ali O, Boudon F, *et al.* (2014) An auxin-mediated shift toward growth isotropy promotes organ formation at the shoot meristem in Arabidopsis. *Curr Biol* **24**: 2335–2342.

Sassi M and Vernoux T. (2013) Auxin and self-organization at the shoot apical meristem. *J Exp Bot* **64**: 2579–2592.

Sattler R. (1988) A dynamic multidimensional approach to floral morphology. In: Leins P, Tucker SC, Endress PK, Cramer J (eds), *Aspects of Floral Development*. Berlin, pp. 1–6.

Sattler R. (1992) Process morphology: structural dynamics in development and evolution. *Can J Bot* **70**: 708–714.

Sauquet H, von Balthazar M, Magallon S, *et al.* (2017) The ancestral flower of angiosperms and its early diversification. *Nature Commun* **8**: 16047.

Scarpella E, Marcos D, Friml J, and Berleth T. (2006) Control of leaf vascular patterning by polar auxin transport. *Genes Dev* **20**: 1015–1027.

Scheres B. (2001) Plant cell identity. The role of position and lineage. *Plant Physiol* **125**: 112–114.

Schimper KF. (1830) Beschreibung des *Symphytum* Zeyheri und seiner zwei deutschen Verwandten der S. *bulborum* Schimper und S. *tuberosum* Jacqu. *Geiger's Magazin für Pharmacie* **29**: 1–92.

Schneeberger R, Tsiantis M, Freeling M, and Langdale JA. (1998) The *rough sheath2* gene negatively regulates homeobox gene expression during maize leaf development. *Development* **125**: 2857–2865.

Schoof H, Lenhard M, Haecker A, *et al.* (2000) The stem population of Arabidopsis shoot meristems is maintained by a regulatory loop between the CLAVATA and WUSHEL genes. *Cell* **100**: 635–644.

Schoute JC. (1913) Beitrage zur Blattstellunglehre. I. Die Theorie. *Recueil de Travaux Botaniques Neerlandais* **10**: 153–339.

Schüpp O. (1914) Wachstum und Formwechsel des Sprossvegetationspruktes der Angiosperm. *Ber Deutsch Bot Ges* **21**: 113–117.

Schwabe W. (1971) Chemical modification of phyllotaxis and its implications. *Symp Soc Exp Biol* **25**: 310–322.

Schwabe W. (1998) The role and importance of vertical spacing at the plant apex in determining the phyllotactic pattern. In: Jean RV, Barabé D (eds), *Symmetry in Plants*. World Scientific, Singapore, pp. 523–535.

Schwabe WW and Clewer AG. (1984) Phyllotaxis — a simple computer model based on the theory of a polarly-translocated inhibitor. *J Theor Biol* **109**: 595–619.

Schwendener S. (1878) *Mechanische Theorie der Blattstellungen*. Engelmann, Leipzig.

Schwendener S. (1883) Zur Theorie der Blattstellungen. Sitzingsber. *Königlich Preussische Akademie der Wissenschaften zu Berlin* **II, S**: 741–773.

Schwendener S. (1909) *Theorie der Blattstellungen. Mecanische Probleme der Botanik*. Engelmann, Leipzig.

Seilacher A. (1991) Self-organizing morphogenetic mechanisms as processors of evolution. Revista Española de Paleontologica, n.° Extraordinaro pp. 5–11.

Sekimura T. (1995) The diversity in shoot morphology of herbaceous plants in relation to solar radiation captured by leaves. *J Theor Biol*. **177**: 289–297.

Selker JML and Lyndon RF. (1996) Leaf initiation and *de novo* pattern formation in the absence of an apical meristem and pre-existing patterned leaves in watercress (*Nasturtium officinale*) axillary explants. *Can J Bot* **74**: 625–641.

Selker JML, Steucek GL, and Green PB. (1992) Biophysical mechanisms for the morphogenetic progressions at the shoot apex. *Dev Biol* **153**: 29–43.

Selvam AM. (1998) Quasicrystalline pattern formation in fluid substrates and phyllotaxis. In: Jean RV, Barabé D (eds), *Symmetry in Plants*. World Scientific, Singapore, pp. 795–809.

Shannon CE and Weaver W. (1949) *The Mathematical Theory of Communication*. University of Illinois Press, Urbana.

Shi B and Vernoux T. (2019) Patterning at the shoot apical meristem and phyllotaxis. *Current Topics in Dev Biol* **131**: 82–107.

Shipman PD. (2010) Discrete and continuous invariance in phyllotactic tillings. *Phys Rev E* **81**: 031905.

Shipman PD and Newell AC. (2004) Phyllotactic patterns of plants. *Phys Rev Lett* **92**(16): 168102-1–168102-4.

Shipman PD and Newell AC. (2005) Polygonal planforms and phyllotaxis on plants. *J Theor Biol* **236**: 154–197.

Shipman PD, Sun Z, Pennybacker M, and Newll AC. (2011) How universal are Fibonacci patterns? *Eur Phys J* **62**: 5–17.

Smith LG, Hake S, and Sylvester AW. (1996) The *tangled-1* mutation alters cell divisions orientations throughout maize leaf development without altering leaf shape. *Development* **122**: 481–489.

Smith RS, Kuhlemeier C, and Prusinkiewicz P. (2006a) Inhibition fields for phyllotactic pattern formation: a simulation study. *Can J Bot* **84**: 1635–1649.

Smith RS, Guyomarc'h S, Mandel T, *et al.* (2006b) A plausible model of phyllotaxis. *Proc Natl Acad Sci (PNAS)* **103**: 1301–1306.

Smyth DR. (2018) Evolution and genetic control of the floral ground plan. *New Phytol* **220**: 70–86.

Snow M and Snow R. (1931) Experiments on phyllotaxis. I. The effect of isolating a primordium. *Phil Trans R Soc London* **221B**: 1–43.

Snow M and Snow R. (1935) Experiments on Phyllotaxis III — diagonal splits through decussate apices. *Philos Trans R Soc London B, Biol Sci* **225**: 63–94.

Snow M and Snow R. (1937) Auxin and leaf formation. *New Phytol* **36**: 1–18.

Snow M and Snow R. (1951) On the presence of tissue tension in apices. *New Phytol* **50**: 185–484.

Snow M and Snow R. (1952) Minimum areas and leaf determination. *Proc R Soc London B* **139**: 545–566.

Snow M and Snow R. (1962) A theory of phyllotaxis based on *Lupinus albus*. *Philos Trans R Soc London B* **244**: 483–513.

Snow R. (1965) The causes of bud eccentricity and the large divergence angles between leaves in Cucurbitaceae. *Philos Trans R Soc London B* **250**: 53–77.

Sokoloff DM, Remizoma MV, Bateman RM, and Rudall PJ. (2018) Was the ancestral angiosperm flower whorled throughout? *Amer J Bot* **105**: 5–15.

Staedler YM and Endress PK. (2009) Diversity and lability of floral phyllotaxis in the pluricarpellate families of core Laurales (Gomortegaceae, Atherospermataceae, Siparunaceae, Monimiaceae). *Int J Plant Sci* **170**: 522–550.

Steele CR. (2000) Shell stability related to pattern formation in plants. *J Appl Mech* **67**: 237–247.

Steeves TA and Sawhney VK. (2017) *Essentials of Developmental Plant Anatomy*. Oxford University Press, New York.

Steeves TA and Sussex IM. (1989) *Patterns in Plant Development*. Cambridge University Press, Cambridge.

Steucek GL, Selker JL, and Reif WE. (1992) Architecture of the plant shoot apex. In: Reiner R (ed), *Natürliche Konstruktionen*. Kurz and Co., Stuttgardt.

Stieger PA, Reinhardt D, and Kuhlemeier C. (2002) The auxin influx carrier is essential for correct leaf positioning. *Plant J* **32**: 509–517.

Stoma S, Lucas M, Chopard J, *et al.* (2008) Flux-based transport enhancement as a plausible unifying mechanism for auxin transport in meristem development. *PLOS Comput Biol* **4**(10): e1000207. DOI:10.1371/journal.pcbi.1000207.

Strickland C, Pearson DA, and Shipman PD. (2016) Formation of square lattices in coupled pattern-forming systems. *Biomathematics* **5**: 1612181. DOI:10.11145/j.biomath.2016.12.181.

Stuurman J, Jäggi F, and Kuhlemeier C. (2002) Shoot meristem maintenance is controlled by a GRAS-gene mediated signal from differentiating cells. *Genes Dev* **16**: 2213–2218.

Swinton J, Ochu E, and MSI Turing's Sunflower Consortium. (2016) Novel Fibonacci and non-Fibonacci structure in the sunflower: results of a citizen science experiment. *Royal Soc Open Sci* **3**: 16009. DOI:10.1098/rsos.160091.

Swinton J. (2005) Watching the daisies grow: turing and Fibonacci phyllotaxis. In: Teuscher C (ed), *Alan Turing: Life and Legacy of a Great Tthinker*. Springer, New York, pp. 477–498.

Swinton, J. (2012) *The Fundamental Theorem of Phyllotaxis Revisited*. http://arxiv.org/abs/1201.1641.

Szpak M and Zagórska-Marek B. (2011) Phyllotaxis instability — exploring the depths of first space. *Acta Soc Bot Pol* **80**: 279–284.

Tait PG. (1872) On phyllotaxis. *Proc R Soc Edinburgh* **7**: 391–394.

Takaki R, Ogiso Y, Hayashi M, and Katsu A. (2003) Simulations of sunflower spirals and Fibonacci numbers. *Forma* **18**: 295–305.

Tamaoki M, Nishimura A, Aida M, *et al.* (1999) Transgenic tobacco overexpressing a homeobox gene shows a developmental interaction between leaf morphogenesis and phyllotaxy. *Plant Cell Physiol* **40**: 657–667.

Tamura Y, Kitano H, Satoh H, and Nagato, Y. (1992) A gene profoundly affecting shoot organization in the early phase of rice development. *Plant Sci.* **82**: 91–99.

Thornley JHM. (1975) Phyllotaxis I. A mechanistic model. *Ann Bot* **39**: 491–507.

Tooke F and Battey N. (2003) Models of shoot apical meristem function. *New Phytol* **159**: 37–52.

Traas J. (2019) Organogenesis at the shoot apical meristem. *Plants* **8**: 6. DOI:10.3390/plants8010006.

Tucker SC. (1961) Phyllotaxis and vascular organisation of the carpels of *Michelia fuscata. Amer J Bot* **48**: 60–71.

Turing AM. (1952) The chemical basis of morphogenesis. *Phil Trans R Soc Ser B, London* **237**: 37–52.

Turing AM. (1992) *Collected Works of A.M. Turing: Morphogenesis*. In: Saunders PT (ed), North-Holland, Amsterdam.

Uyttewaal M, Burian A, Alim K, *et al.* (2012) Mechanical stress acts via katanin to amplify differences in growth rate between adjacent cells in *Arabidopis. Cell* **149**: 439–451.

Vakarelov I. (1998) Changes in phyllotactic pattern structure in *Pinus* due to changes in altitude. In: Jean RV, Barabé D (eds), *Symmetry in Plants*. World Scientific, Singapore, pp. 213–230.

Valladares F and Brites D. (2004) Leaf phyllotaxis: does it really affect light capture? *Plant Ecol* **174**: 11–17.

Valladares F, Skillman JB, and Pearcy RW. (2002) Convergence in light capture efficiencies among tropical forest understory plants with contrasting crown architectures: a case of morphological compensation. *Amer J Bot* **89**: 1275–1284.

Van Berkel K, de Boer RJ, Sheres B, and ten Tusscher K. (2013) Polar auxin transport: models and mechanisms. *Development* **140**: 2253–2268.

Van der Linden FMJ. (1998) Creating phyllotaxis from seed to flower. In: Jean RV, Barabé D (eds), *Symmetry in Plants*. World Scientific, Singapore, pp. 487–521.

Van Iterson G. (1907) Mathematische und Mikroskopisch-Anatomische Studien über Blattstellungen nebst Betraschtungen über den Schalenbau der Miliolinen. Gustav Fischer, Jena.

Van Iterson G. (1964) Nieuwe Studiën over Bladstanden. I. Verhandelingen der koninklijke nederlandse Akademie van Wetenschappen. *Afd. Naturkunde* **52**(2): 1–151.

Van Mourik S, Kaufman K, van Dijk ADJ, *et al.* (2012) Simulation of organ patterning on the floral meristem using a polar auxin transport model. *PLOS One* **7**: e28762.

Veen AH and Lindenmayer A. (1977) Diffusion mechanism for phyllotaxis, theoretical, physic-chemical and computer study. *Plant Physiol* **60**: 127–139.

Veit B, Briggs SP, Schmidt RJ, *et al.* (1998) Regulation of leaf initiation by *terminal ear1* gene of maize. *Nature* **393**: 166–168.

Vernoux T, Brunoud G, Farcot E, *et al.* (2011) The auxin signalling network translates dynamic input into robust patterning at the shoot apex. *Mol Syst Biol* **7**: 508. DOI:10.1038/msb.2011.39.

Vernoux T, Kronenberger J, Greandjean O, *et al.* (2000) *PIN-FORMED1* regulates cell fate at the periphery of the shoot apical meristem. *Development* **127**: 5157–5165.

Vieth J. (1998) Biastrepsis and allomery of stems of *Dipsacus sylvestris* Mill. In: Jean, RV, Barabé D (eds), *Symmetry in Plants*. World Scientific, Singapore, pp. 359–392.

Waddington CW. (1940) *Organizers and Genes*. Cambridge University Press, Cambridge.

Waites R, Selvadurai HRN, Oliver IR, and Hudson A. (1998) The *phantastica* gene encodes a MYB transcription factor involved in growth and dorsoventrality of lateral organs in *Antirrhinum*. *Cell* **93**: 779–789.

Wang Y and Li J. (2008) Molecular basis of plant architecture. *Annu Rev Plant Biol* **59**: 253–279.

Wardlaw CW. (1949a) Experiments on organogenesis in ferns. *Growth* **13**(suppl.): 93–131.

Wardlaw CW. (1949b) Experimental and analytical studies of pteridophytes. XIV. Leaf formation and phyllotaxis in *Dryopteris aristata* Druce. *Ann Bot* **13**: 164–198.

Weigel D, Alvarez J, Smyth DR, *et al.* (1992) LEAFY controls floral meristem identity in *Arabidopsis*. *Cell* **69**: 843–859.

Wiesner J. (1875) Bemerkungen über Rationale und Irrationale Divergenzen. *Flora* **LVIII**: 113–115, 139–143.

Wildoer JW, Venema LC, Rinzler AG, *et al.* (1998) Electronic structure of atomically resolved carbon nanotubes. *Nature* **391**: 59–62.

Williams RF. (1975) *The Shoot Apex and Leaf Growth*. Cambridge University Press, Cambridge.

Wilson PI. (1995) The geometry of golden section phyllotaxis. *J Theor Biol* **177**: 315–323.

Wyrzykowska J, Pien S, Shen WH, and Fleming, A. (2002) Manipulation of leaf shape by modulation of cell division. *Development* **129**: 957–964.

Yagil G. (1985) On the structural complexity of simple biosystems. *J Theor Biol* **112**: 1–23.

Yamada H, Tanaka R, and Nakagaki T. (2004) Sequences of symmetry-breaking in phyllotactic transitions. *Bull Math Biol* **66**: 779–789.

Yang F, Bui HT, Pauler M, *et al.* (2015) A maize glutaredoxin gene, abphyl2, regulates shoot meristem size and phyllotaxy. *Plant Cell* **27**: 121–131.

Yang Y, Hammes UZ, Taylor CG, *et al.* (2006) High-affinity auxin transport by the AUX1 influx carrier protein. *Curr Biol* **16**:1123–1127.

Yeatts FR. (2004) A growth-controlled model of the shape of a sunflower head. *Math Biosci* **187**: 205–221.

Yin X, Lacroix C, and Barabé D. (2011) Phyllotactic transitions in seedlings: the case of *Thuja occidentalis*. *Botany* **89**: 387–396.

Yin X and Meicenheimer RD. (2017) The ontogeny, phyllotactic diversity, and discontinuous transitions of *Diphasiastrum digitatum* (Lycopodiaceae). *Amer J Bot* **104**: 1–16.

Yonekura T, Iwamoto A, Fujita H, and Sugiyama M. (2019) Mathematical model studies of the comprehensive generation of major and minor phyllotactic patterns in plants with a predominant focus on orixate phyllotaxis. *PLoS Comput Biol* **15**: e1007044.

Yoshikawa HN, Mathis C, Maïssa P, *et al.* (2010) Pattern formation in bubbles emerging periodically from a liquid free surface. *Eur Phys J E* **33**: 11–18.

Yotsumoto A. (1993) A diffusion model for phyllotaxis. *J Theor Biol* **162**: 131–151.

Young DA. (1978) On the diffusion theory of phyllotaxis. *J Theor Biol* **71**: 421–432.

Zagórska-Marek B. (1985) Phyllotactic patterns and transitions in *Abies balsamea*. *Can J Bot* **63**: 1844–1854.

Zagórska-Marek B. (1987) Phyllotaxis triangular unit: phyllotactic transitions as the consequences of the apical wedge disclinations in a crystal-like pattern of the units. *Acta Soc Bot Pol* **56**: 229–255.

Zagórska-Marek B. (1994) Phyllotactic diversity in *Magnolia* flowers. *Acta Soc Bot Pol* **63**: 117–137.

Zagórska-Marek B and Szpak M. (2008) Virtual phyllotaxis and real plant model cases. *Funct Plant Biol* **35**: 1025–1033.

Zagórska-Marek B and Szpak M. (2016) The significance of γ- and λ-dislocations in transient states of phyllotaxis: how to get more from less — sometimes! *Acta Soc Bot Pol* **85**: 3532. DOI:10.5586/asbp.3532.

Zagórska-Marek B and Wiss D. (2003) Dislocations in the repetitive unit patterns of biological systems. In: Nation J, Trofimova I, Rand J and Sulis W (eds), *Formal Descriptions of Developing Systems*. Kluwer Academic Press, Dordrecht, pp. 99–117.

Zotz G, Reichling P, and Valladares F. (2002) A simulation study on the importance of size-related changes in leaf morphology and physiology for carbon gain in an epiphytic bromeliad. *Ann Bot* **90**: 437–443.

Zoulias N, Duttke SH, Garcês H, *et al.* (2019) The role of auxin in the pattern formation of the Asteraceae flower head (capitulum). *Plant Physiol* **179**: 391–401.

Index